GOOD AND FAITHFUL LABOR

Recent titles in
Contributions in American History
Series Editor: Jon L. Wakelyn

GOOD AND FAITHFUL LABOR

FROM SLAVERY TO SHARECROPPING IN THE NATCHEZ DISTRICT, 1860-1890

• Ronald L. F. Davis •

CONTRIBUTIONS IN AMERICAN HISTORY, NUMBER 100

Greenwood Press

WESTPORT, CONNECTICUT • LONDON, ENGLAND

Copyright Acknowledgements

Versions of chapters 3, 5, and 7 have appeared in the *Journal of Negro History* (January 1977): 60-81; in *Essays in Economic and Business History*, ed. James H. Soltow (East Lansing, Mich.: Michigan University Press, 1979); and in *From the Old South to the New*, ed. Walter J. Fraser, Jr. and Winfred B. Moore, Jr. (Westport, Conn.: Greenwood Press, 1981). They are used here by permission.

Library of Congress Cataloging in Publication Data

Davis, Ronald L. F.
    Good and faithful labor.

    (Contributions in American history, ISSN 0084-9291;
no. 100)
    Bibliography: p.
    Includes index.
    1. Share-cropping—History. 2. Afro-Americans—
Mississippi—Adams County—Economic conditions. 3. Afro-
Americans—Louisiana—Concordia Parish—Economic con-
ditions. 4. Concordia Parish (La.)—Rural conditons.
5. Adams County (Miss.)—Rural conditions. I. Title.
II. Series.
HD1478.U6D38   306'.3   81-13367
ISBN 0-313-23134-6 (lib. bdg.)   AACR2

Library of Congress Catalog Card Number: 81-13367
ISBN: 0-313-23134-6
ISSN: 0084-9291

First published in 1982

Greenwood Press
A division of Congressional Information Service, Inc.
88 Post Road West
Westport, Connecticut 06881

Printed in the United States of America

10 9 8 7 6 5 4 3 2 1

To my wife, Patricia, and my children, Stacey and Christian, who shared so much of themselves with me in the creation of this book.

CONTENTS

ILLUSTRATIONS

TABLES

ACKNOWLEDGMENTS

This project was begun at the University of Missouri in Columbia, and I would like to thank Harvey Becher, Hal Christensen, H. Roger Grant, Harry Holmes, Richard McLeod, Thomas Kelley, Peter Romanofsky, and Jerome Steffen, who made those long ago years of initial research so pleasant and rewarding. Several of these friends (Hal and Tom) even helped me with some of the early collection of materials. In the years since leaving Missouri, I have taught at California State University, Northridge. Again, I have had the good fortune to be surrounded by a fine group of fellow historians, some of whom contributed directly to the work. A word of thanks is due to John Broesamle, Allan Dirrim, Sheldon Harris, Paul Koistinen, Michael Meyer, Leonard Pitt, Ron Schaffer, and Morris Schonbach for their encouragement over the years. Reba Soffer contributed her critical reading of portions of the manuscript, and the work is much stronger because of her interest in it. Among those who helped directly is my warm friend and colleague, Shiva Bajpai. Shiva has provided the kind of encouragement and support possible only when one's good friend is also a profound scholar. I am especially thankful to him for sharing his understanding of the economy and practices of traditional agriculture.

I also want to express my appreciation to the many librarians, archivists, and municipal, state, county, and parish officials who aided my research to the fullest. My work took me to the Louisiana State University and to the Louisiana Records Commission in Baton Rouge; to the City Halls of Records in Vidalia, Louisiana, and in Natchez, Mississippi; to the Library of Congress and to the National Archives; to the

Mississippi Department of Archives and History in Jackson; to the Baker Library in Cambridge; to Tulane University; and to numerous bus stops and byroads throughout the Natchez District. All of the officials and personnel whom I met along the way have my sincere thanks for their help.

I owe a special thanks to a group of research assistants and students who contributed greatly to this project, especially Chapters 5 and 7. In many ways, this part of the task was one of the most enjoyable, and I learned far more than they from the experience. These students searched the manuscript census records, and typed, coded, discussed, evaluated, argued, and shared with me their important insights. They were Donna Krug, Nancy Sullivan, Francine Bloom, Elfrieda Stone, Kim Paul, Claire Bleiman, and Margaret Moe.

With regards to the women who managed to type the many copies of this manuscript over the years, I can only hope that this brief acknowledgment will serve as a token of my regard for them. Department secretaries Claire Bleiman, Marcia Dunnicliffe, and Nancy Meadows did the yeoman job of a chapter here and an essay there, usually with no idea of what would become of their work. Joyce Gerritsen, Donna Krug, and Jan Wasson typed the manuscript at three different moments in its creation. Their work went far beyond the call of duty.

Among the readers who improved the manuscript, I owe a special debt of appreciation to LaWanda Cox, Stanley Engerman, and several anonymous readers. Professors Cox and Engerman were painstakingly thorough in their criticisms and wonderfully encouraging in their support. Others who offered valuable comments and criticisms of parts of the manuscript include Louis S. Gerteis, George W. Hopkins, Melton A. McLaurin, John A. Tomaske, Jonathan M. Weiner, and Gavin Wright.

Many other people have given valuable assistance at various stages of the book's progress. Especially helpful was the editor of the series to which this study belongs, Jon L. Wakelyn. I would also like to thank Margaret M. Brezicki, Anne Kugielsky, and Betty Pessagno of Greenwood Press for their excellent editorial assistance in the preparation of this book.

But the man to whom I am especially indebted is my mentor, Harold D. Woodman. It was he who first suggested the topic (or at least an aspect of it), introduced me to research in southern archives, guided its development, taught me whatever I know about economic history, and served as a model of what the historian should be as a scholar and a friend.

I owe my greatest debt to Patricia, my wife and best friend. Indeed, this study should include her name on the title page in view of her countless hours of coding and reading—all performed while either working as a teacher or in research or, lately, in raising a family.

I would also like to thank the Council on Research in Economic History for the award of an Arthur H. Cole Grant for research and the Foundation of the California State University at Northridge for its help in funding my research, travel, and microfilming expenses.

GOOD AND
FAITHFUL LABOR

• 1 •

INTRODUCTION

Slavery died with the Civil War, but postbellum southern society did not lose its antebellum slave character. The most visible evidence of this was the entrapped and deprived condition of black people as sharecroppers and tenants. While slavery had indeed vanished, the plantation system of agriculture had survived. White landlords remained the South's dominant class, and black laborers in southern farming were still, as they had been in slavery, a dependent and impoverished caste of racially oppressed people. The reason for this continuity rested largely with the institution of sharecropping and its "peculiar" features.

Although much has been written about sharecropping, notably its recent sociology and its so-called reasonableness as an economic system, nowhere is there an adequate explanation for its historical appearance. It seems simply to have emerged out of the ashes of the Civil War, as if slavery's roots were too deeply implanted to be destroyed by the mere granting of freedom to southern blacks. Once it took hold as the main way of living and working for southern people, sharecropping proved to be a profitable and effective way of growing cotton in a largely preindustrial and extremely racist society. But its viability as an economic system does not fully explain its origins.

It is proposed here that sharecropping originated because southern blacks embraced it as a means of escaping a wage system of labor which they found to be slavery in all but name. The emancipated slaves brought with them into freedom an overwhelming desire for independence in their daily work and lives. During the last days of the Civil War, they had hoped to become their own masters as landowning farmers. But when the victorious North refused to break up the plantations into homesteads for the former slaves, the South's freedmen were left with few resources except their labor and their intense desire for freedom. How, they asked, by their actions and in the ways of an impoverished, rightless, and illiterate mass of people, could their newly won freedom be preserved? The answer, they eventually came to believe, was in sharecropping.

Although there were some similarities, the freedman's reaction to emancipation differed from that of the land-attached, Western European peasant. For one thing, when the peasants became free laborers, they frequently resisted becoming cash laborers or "kind" workers, finding it a less secure way of life than their earlier land-attached status. The liberty to be hired and fired was not a freedom they accepted easily or with much optimism. In contrast, few former slaves regarded their slave dependency with affection or thought it a better way of living than that provided by freedom. The slave's dependency had been devoid of any rights, privileges, security, or traditional protections except those achieved by chance in return for docile behavior or through the master's compassion. Unlike the peasant, whose attachment to the land made him somewhat independent of his landlords and squires, the slave's legal bondage as personal property rendered him subject to his master's whim and fancy, to the accidents of time and place, and to the fluctuations of the business cycle. Even those slaves with kind and paternal masters well understood the degree of their insecurity. Thus conditioned in slavery, what freedmen wanted most was self-direction in their work, the security of their families, and the freedom to be unattached to persons and places other than themselves and their own farms. Unlike a displaced peasantry, freedmen saw the insecurity of freedom in the context of what they had known in slavery.

Their former masters, on the other hand, wished to recreate an economic and social system wherein blacks would be attached, if not to a white master as chattel property, at least to the plantation as a controlled, stable, and subservient class of low-paid laborers. Immediately after the war, planters tried to achieve this new system through local legislation (Black Codes) that denied blacks the right to be unemployed or to hold their labor off the market, defining those that were or did as vagrants subject to arrest and work gangs. This plan for controlling blacks, although never fully implemented thanks to the intervention of the Freedmen's Bureau, satisfied two heartfelt convictions of the old planter class: first, the belief that the former slaves were incapable of working as self-directed laborers and, second, the certainty that blacks, for the good of all concerned, must remain a controlled people.

Others, including northern abolitionists, U.S. Army personnel, agents of the Freedmen's Bureau, and even Yankee businessmen, sought to introduce into southern agriculture their own free labor attitudes toward life and the practice of farming. They thought it reasonable and just for southern blacks to sell their labor to the highest bidders, to work at their jobs long and hard and well as disciplined wage hands, and thereby to accumulate enough capital from their earnings to buy small family farms of their own. In this way, yeoman blacks would become truly independent of the antebellum planter class. Forced to haggle and bargain for their wages in a free economy, the former slaves would soon develop, insofar as each individual was able, the one essential requirement of the free labor system: self-motivated ambition.

Of the competing views mentioned above, the one that least succeeded was that held by the North, mainly because it depended on an expanding market for the product of black labor. Had cotton prices remained high for a half-century after slavery, freedmen might have obtained high wages, accumulated capital, and achieved a free farming status, or at least migration out of the southern economy and society as planters adopted labor-saving technology to offset the high wages paid. This scenario assumes, of course, a strong interest by northern businessmen in protecting the competitiveness of black labor as well as the absence of any major obstacles to black migration. The downward spiral of cotton prices that did occur, together with those forces opposed to the emergence of a free labor system of farming (poverty and racism alike), destroyed any possibility of a free labor market system of agriculture in the postbellum South.

What emerged instead was an unusual form of farming that involved neither the cash nexus of wages between labor and capital nor the landed security of a peasant tenancy. Blacks, rather, worked in family groups, paying a share of their produce as rent or receiving a share of the crops made as their pay, depending on their relative assets. In both cases, they were impoverished, with few legal rights or traditional privileges, and tied to a system of supply, though not to a particular master or place. While their former masters had all but disappeared as resident managers of a plantation estate or labor force, they often retained their properties as landlords and absentee capitalists. In their place arose a new breed of plantation businessman, the merchant, who controlled the black sharecropper's provisions and thus his life.

But sharecropping was not a compromise between planters and freedmen and Yankees. Of all of the actors involved in the system's birth, only the freedmen wanted it. Why they wanted sharecropping, why they were able to obtain it, and how it became something other than what they imagined it would be is the subject of this study.

THE LITERATURE

The scholarship on sharecropping is surprisingly rich despite the neglect of its origins. Almost from its beginnings as the South's dominant system of postbellum agriculture, sharecropping has been the object of scrutiny. In the 1880s and 1890s, proponents of the "New South" attacked the system as the chief impediment to the region's economic progress. In the twentieth century, professional sociologists, anthropologists, and government-funded researchers produced monumental studies of the system's caste and class structure. Their work was apparently so exhaustive that the noted Swedish sociologist, Gunnar Myrdal, included only a summary allusion to the subject in his classic work on race, *The American Dilemma*. Recently, as a result of a pathbreaking study of the theory of tenancy by economist Steven Cheung and the progression of interest among historical economists from the study of slavery to its aftermath, the subject is once again under study.[1]

Over the years, scholars have, in passing, speculated on the elements involved in the birth of sharecropping. In general, the studies have emphasized (1) the planter's dilemma in working inefficient laborers, (2) the freedman's role in striving to carve out a meaningful place for himself in southern agriculture, (3) the impact of sustained agricultural depression which forced planters and freedmen to compromise their positions and accept sharecropping, and (4) the influence of the varied costs and risks of farming, especially those of supervision, in the tenure choices of wages, sharecropping, and fixed rent tenancy. Scholars have acknowledged the role of all of these factors but have not generally given them equal weight.

According to the first approach, planters introduced sharecropping after their experiments with freedmen as fixed wage laborers had ended in failure. The planters made the change to better protect their investments by tying the freedman's earnings to the harvest instead of paying him wages set early in the year at the time of the contract. Believing that they had found a means of forcing labor to share in the losses stemming from their irresponsibility as free workers, southern whites initiated the share system as a temporary arrangement until the day northern soldiers left southern planters to their own affairs.[2]

Other studies have suggested that freedmen played a crucial role in the emergence of sharecropping. Once blacks realized that freedom meant working for their old masters, they began withholding their labor to pressure planters into making concessions in the working conditions and terms of labor. Work slowdowns, insubordination, and refusals to contract won for the freedmen the right to cultivate gardens, to work six-day weeks, to own and market chickens and pigs, and, most importantly, to work somewhat under their own direction and in their own time.[3] Although the literature fails to explain how all of this resulted in sharecropping, it does suggest that the system's adoption was related to the actions of labor.

Still others stress the sustained depression in southern agriculture as the main reason for the emergence of sharecropping. Bankrupt planters, unable to obtain much cash or long-term credits, adopted sharecropping as the only way to continue planting. In turn, widespread crop failures, the extensive use of the crop lien, and the falling price of cotton in the late 1860s institutionalized sharecropping as a form of tenancy and wage settlement. The mechanism turned planters into merchants as they leased their lands and moved to town to conduct their supply business. In turn the merchants were transformed into planter-landlords as they foreclosed on lands mortgaged for plantation supplies. The business of supply thus became linked to land and labor through sharecropping and fixed rent tenancy.[4]

Finally, recent works emphasize that planters and laborers chose sharecropping because it guaranteed mutual risk-sharing and aggregate risk reduction, lowered supervision costs, and opened to freedmen alternative means of income as they were free to shift their efforts between agriculture, fishing, hunting, and other remunerative activities. Where tenancy prevailed, the theory continues, the choice probably reflected the high costs of supervising large groups of freedmen on the larger plantations.[5]

Unfortunately, because none of these four approaches addresses the issue of how sharecropping originated directly, there is little consensus on the matter, and the merits of each are often flawed by untested assumptions and the lack of historical evidence. Among the most recent students of sharecropping—the econometricians (or historical economists)—problems of methodology and conceptualization characterize their findings and generally distort their history. Being narrowly economic, they generally neglect so-called noneconomic evidence while assuming that the economic rationality of a system is explanation enough for its cause and development over time.[6]

Examples of these problems may be found in the recent work of two leading historical economists, Joseph D. Reid and Robert Higgs. Reid, in a perceptive essay, maintains that sharecropping emerged as a kind of compromise between planters and freedmen in a calculated response to the existing market and to price conditions for cotton. In the bargain that was struck, blacks gained some independence and planters obtained laborers while still remaining in control of the crucial decisions of crop management. The system was attractive to both parties mainly because they saw the possibilities of sharing with one another the risks of production. Although Reid's reasoning is incisive, it fails to provide a satisfying historical explanation because so much is left out of its formulation. Among other things unresolved by this view are the role of land hunger among freedmen, racism, the goals of the Freedmen's Bureau, and the relationship between the freedman's work ethic, whatever it was, and the emergence of sharecropping.[7]

Similarly, Higgs tells us that the move to sharecropping reflected primarily the demands of simple cost accounting. Planters and landlords found that sharecropping involved lower supervision costs, depending on the size of the place. In citing the example of a planter in 1901 who told an investigative body that he pursued the "iniquitous" system of sharecropping because he could "not get the labor to work otherwise" and because it was a "more" profitable scheme "under the circumstances," Higgs concludes that planters adopted the system because it was the most "remunerative alternative" available.[8] Yet, the planter's own words suggest that the essential determining factor was the refusal of his black laborers to work under any other system. Why this was the reality is an historical question that cannot be explained solely in terms of economic incentives. And that planters made money "under the circumstances" may have been a description of their reality rather than an interpretation of its cause. Slave masters had also made money with slavery, but such profits only partly explain their commitment to the system.[9]

THE FAITHFUL WORKER

One thing is certain: few southern whites believed that the emancipated blacks were capable of self-directed labor. According to the planter's traditional view of labor under slavery, the faithful worker (the slave totally subservient to the planter's will) had been the efficient worker. The very survival of slavery as a system of plantation agriculture had required this definition of labor efficiency. Outnumbered as they were, antebellum planters valued the docile slave above the independent slave. The master desired both productivity and subordination among slave hands, but the system placed the highest premium on subordination. When the choice was made as to who would be sold and who would stay, the docile hand usually remained. Planters knew that the subservient slave would follow orders and produce, whereas the hand who produced without following orders, or in defiance of orders, threatened the essential character of plantation slavery: low-skilled, gang labor and well-coordinated farming.

A glance at the records of employer complaints filed with the Freedmen's Bureau after the Civil War indicates that most southern planters continued to measure a worker's value in terms of faithfulness and subservience. In the hundreds of wage contracts signed between planters and freedmen, the most common stipulation demanded by planters was the worker's promise to provide "good and faithful labor." It was this promise that planters most often accused their hired hands of breaking. Refusing and failing to do the countless little things on the plantation that had been commonly done by slaves, freedmen resisted subservience and thus proved, from the planter's point of view, their unreliability and inefficiency as free workers. Among the examples cited were refusing to run errands in town, to haul wood, to work on the levees, to keep gardens for the landlord's use, to repair fences, and to drive the employer's wagon.[10] Most intolerable, however, was the sight of black workers, hired as wage laborers, dropping their hoes and leaving the fields at will to chase rabbits or go to the well for a drink or simply to take an unscheduled rest.[11] Such unfaithful labor convinced planters that they need have little compunction in breaking the remaining terms of the contracts signed.

This tendency to view labor in terms of its docility and subordination was aptly expressed by a contributor to *The American Farmer* in 1868. After explaining that it was doubtful that the "poor freedman will ever become reliable enough as a laborer, or sufficiently trustworthy in a moral point of view, to justify his becoming a vital element in the permanent establishment of our agricultural prosperity," "Vox," as the author signed himself, hastened to add that the difficulty with hired white labor was equally insurmountable. "Unlike 'Cuffy,' the poorer class of white men suffer from an indisposition on their part to regard anyone as master and a reluctance to receive orders except in the form of polite suggestions." While blacks were hopeless as long as they remained free, according to "Vox," they could at least function as directed labor under conditions of strict control. But yeoman whites, whom the former slave masters had not enslaved, were ineffective as farm laborers because they were simply too independent. Nor did "Vox" see much hope in introducing "a swarm of miserable Chinese," since they would only "add to that poisonous undercurrent of immorality and licentiousness" which he feared "would be long present in the South as a result of slavery's end." Blacks and other nonwhites, being racially tainted, were unable to work without supervision regardless of their training, and southern poor whites were unprepared to work as disciplined laborers. In the end, "Vox's" only solution, since slavery seemed out of the question, was to devise some means of educating the South's yeoman whites to accept direction and authority. If this could be accomplished, he concluded, then the South would obtain something never found in slavery: intelligent labor.[12]

The point, of course, is that neither "Vox" nor many other southern planters believed self-directed labor a possibility for either freedmen or poor whites. And it was not only that they were racists, but also they were accustomed to being masters.

What about the work habits of blacks in the immediate postwar period? They had hardly lost their ability to labor long and hard when they became free men, and their frequent turning from the fields to hunt or fish may have reflected the actual requirements of their tasks in cotton rather than a weakened work ethic. Yet, the charges against them by their former masters as unreliable and inefficient workers are too numerous to dismiss so easily. If slavery had produced a master class unable to think of laborers except as servile and dependent workers, then slavery may have also produced a class of laborers unable to function as punctual, efficient, sober, disciplined, and wage-motivated employees.[13] Students of working-class history have shown how common it was for preindustrial people to resist the discipline and control demanded by industrialism. The transition from rural farm life, village living, and craft artisanship to the status of a modern working class was a demoniac process in which one way of working and living gave way reluctantly to another.[14] The disciplined, and often guilt-ridden, work ethic of modern society was characterized at birth by the lateness, absenteeism, drunkenness, restlessness, weak or nonexistent wage responsiveness, and general unreliability or inefficiency of labor— the very characteristics "Vox" found among southern rural folk both black and white.

Another important consideration is the "culture of poverty" analysis of much of modern sociology which would find the so-called irresponsibility of the emancipated slaves a not unexpected result of the painful experience of slavery. This is not the thesis of genetic or biological impairment as much as it is the theory that slavery may have predisposed southern blacks to apathy, manifested by cynicism and despair, and a refusal to believe that much could be accomplished in life.[15] William Faulkner touches on a similar idea in his black characters of Ringo (*The Unvanquished*) and Joe Christmas (*Light in August*) and especially by making his barn-burners embittered white sharecroppers but never blacks.[16]

But neither the point about the casual work discipline of earlier times nor the "culture of poverty" analysis are on the mark. Unlike the new working class of incipient industrialism, freedmen were asked to conform to a way of working and living that they had previously known. That many resisted is well documented, but it is not clear how many or why. Their failure to accept a modified slavery does not indicate that slavery had robbed them of a work ethic or that they were unprepared to function as free, wage-motivated workers.[17] Moreover, the fact that their eventual compliance was accomplished by terror rather than by wage incentives, by debt peonage and illiteracy rather than by job and status mobility, and by poverty and underemployment rather than by family farms or full employment suggests that any noted apathy on the part of blacks may have been as much a result of the conditions of their freedom as it was a legacy of slavery. Indeed, the notion of a culturally inherited character of despair conflicts markedly with what we know of the vibrant slave community that had evolved through the cracks and interstices of the system—a community that was both self-reliant and culturally intact. Such a notion also ignores the optimism and the daring testing of their freedom at the first sight of northern troops by refugee blacks from the Sea Islands to Vicksburg. Finally, it denies the industriousness of black sharecroppers and tenants for five generations after slavery. Still, what must be asked is on what terms were freedmen willing to work and planters willing to hire, and how did these terms result in sharecropping?

THE FREEDMAN'S RELATIVE EFFICIENCY

Both contemporary observers and later historians agree that sharecropping, once it took hold, was a sorry way to run a plantation. Much of this opinion, however, was and is misinformed insofar as it refers to the relative, personal efficiency of blacks as farmers and workers. In 1913, U. B. Phillips told students at the University of Virginia that the efficiency of blacks in cotton production had declined by 35 percent since 1860:

To compare Negro efficiency in cotton production before and since the War, it is necessary to select districts where no great economic change has occurred except the abolition of slavery—where there has been no large introduction of commercial fertilizers for example and no great ravages by boll weevil. A typical area for our purpose is the Yazoo Delta in Northwestern Mississippi. In four typical counties there—Tunica, Coahonia, Bolivar and Issaguena—in which the Negro population numbers about ninety percent of the whole, the per capita output of cotton was two and one-third bales of five hundred pounds each in slavery, while in 1910 and other average recent years it was only one and one-half bales per capita. That is to say, the efficiency of the Negroes has declined thirty-five percent.[18]

According to Phillips, the freedman's poor comparative showing was caused by the breakdown of the antebellum plantation system and the subsequent freeing of blacks from the supervision of white managers.

Although slavery had forced planters to undercultivate their farms to the picking capacity of their slaves, the plantation system had nevertheless enabled them to supervise crude labor at routine tasks with better results than was possible on smaller farms.[19] But sharecropping, in Phillips's opinion, by allowing blacks to contract in family units and to work generally unsupervised, undermined the antebellum plantation's systematic, efficient organization of labor.[20] To make matters worse, sharecropping continued the South's antebellum disadvantage of cultivating cotton with dependent labor, since freedmen contracted in families on a yearly basis and were generally linked to the place by the bonds of sentiment, sympathy, and debt.[21]

Since 1913, scholars have generally endorsed Phillips's thesis except that they emphasize the efficiency of the large antebellum plantation even more than he did. In this view, the slave plantation achieved economic advantages in marketing, management, and finance that were unavailable to smaller farmers before the war and, by implication, to freedmen sharecroppers and tenants after slavery.[22] This position is well summarized by C. Vann Woodward's assertion that the postbellum system of farming operated "minus such scant efficiency, planning, responsible supervision, and soil conservation as the old system provided."[23] Two historical economists, Stanley L. Engerman and Robert W. Fogel, agree with this view and have added considerable support to the suggestion that economies of scale probably occurred in the slave sector of southern agriculture.[24]

But the Phillips tradition is unconvincing in regard both to the comparative efficiency of blacks and to the relative efficiency of the large slave plantation. Much of the argument for economies of scale (savings associated with the size of the operation) on the slave plantation rests on certain unwarranted assumptions about the buying and selling of cotton. According to this thesis, the wealthiest planters purchased in quantity at discount prices and received the top dollar possible for their crops, thus achieving savings that partially justified the size of their operations. Or, in terms of the market, large planters enjoyed advantages because they were able to hold out for or demand higher prices. But studies of the South's marketing system have found little evidence to support this thesis. It seems that a remarkable uniformity operated in prices and interest rates regardless of the size of the plantation.[25]

Even had such advantages existed, there is no reason to believe that any obtained savings affected either the productivity or the personal efficiency of slaves (defined here as meaning how well the slave worked with the tools, land, and capital available). Rather, the sources suggest that profits from any source were plowed into more land and slaves instead of being invested in labor-saving devices, thus leaving the large plantation slave's efficiency unaffected in a comparative sense.[26]

But perhaps the plantation slave enjoyed the advantage of managerial skills or superior organization which were unobtainable on the smaller farms. According to a recently stated view, planters and overseers closely supervised their plantation slaves in specialized task groups amenable to work speedups. The result was an increased productivity or more intensive labor per man-hour than was possible on the smaller farms.[27] Is it realistic, however, to suppose that the large plantation's factory-like division of labor increased the personal efficiency of labor? For the most part, productivity in cotton required diligent hoeing and chopping to keep the crop out of grass, intensive picking before bad weather set in at harvest time, and clean picking, which meant some attention to the job. Although all three of these duties required substantial skill, especially the chopping and clean picking, it is not clear how managerial skills could be better achieved on the large plantation insofar as the tasks of labor were concerned.[28]

Scholars who endorse the view of managerial advantages do so largely on the basis of evidence that planters acted as rational businessmen interested in organizing and working their slaves with such efficiencies in mind.[29] The planters' careful supervision of their labor, however, does not per se constitute proof that they achieved better results than their less systematic neighbors on smaller farms. The larger plantations required a method of organized, careful, and close supervision of slaves partly as a means of maintaining social control and partly as a result of the problems of achieving a minimum degree of order. Farmers on smaller farms, working side by side with their hands in the fields, could abandon gang labor, task assignments, and rigorous supervision without experiencing the insubordination and labor shirking that would have accompanied similar labor practices on the large plantation.

In addition, it must be recognized that the working of a controlled and large labor force in a set routine and with a division of their labor may have contributed little to the actual effectiveness of that labor. Not even those late nineteenth-century industrial factories of the eastern United

States clearly achieved labor efficiency until the advent of scientific management. The major breakthrough of scientific management (Taylorism) in American industry rested upon the recognition that as long as workers controlled any aspect of the job, even with the new machinery capable of increasing the productivity of labor (machinery not available in southern agriculture), workers affected, determined, and most often set the ultimate efficiency levels of their labor. With "Taylorism," the goal of management was to move beyond the simple routinization or division of labor to the removal from the worker of all decisions and conceptual activity in the performance and process of production. No mental contribution by workers, no decisions about the method or procedure of work by workers, and no individualized skill application could be allowed if the full fruits of the division of labor made possible by the newly introduced labor saving devices were to be achieved.[30] Although it is doubtful that slave masters encouraged blacks to contribute to the planning of the work routine on the larger plantations, the preindustrial nature of the work, the blacks' ability to counter labor intensification with carelessness and poor performance under the mask of their so-called stupid and lazy character, and the sheer inability of even the most dutiful managers to supervise rigorously every aspect of the job (no plantation enjoyed the contribution of the industrial engineer) created chronic obstacles to the achievements of an industrial management of a later age.

The greatest advantage in the modern division of labor and its scientific management was unavailable to the antebellum planter: the ability to purchase the precise quantity of skill or labor required—the ability, as Phillips so shrewdly understood, to deal quickly and ruthlessly through firings with antagonistic, shirking, undisciplined, lazy, or incompetent workers.[31] While slave masters often purchased slaves with a concern for labor skills and the number needed for the job, few sold their slaves with this concern in mind. Not only were slaves valuable in and of themselves as a commodity, but they were also an attached people capable of emotional bonds to the master, the plantation, and to one another. These bonds were not easily broken, even though they afforded no real security for the slaves. As a result, the organization of the typical slave plantation reflected the peculiar characteristics of the slave society involved. It was a complicated system indeed in the case of those planters who considered their plantation enterprises as home, estate, and family.

This is not to say that there were no advantages associated with the size of a plantation. To name only a few, large planters could better afford ginning equipment, wagons, and other farm implements than their smaller neighbors, thus possibly increasing the productivity of their workers. In addition, as indicated by Lewis C. Gray, the death of slaves on the larger plantations meant a relatively smaller share of capital lost. And being wealthier, these slave masters could better afford the best lands.[32]

Such advantages, although significant, were usually offset by other problems. Gray also notes that the larger planters frequently worked more acres than their hands could efficiently cultivate.[33] Cotton picked late in the season was of marginal quality and was often poorly handled in the planter's eagerness to get as much land into cotton as possible. Because of overplanting, the large planter could not work the land as well as his smaller neighbors, and so actual crop yields per acre might be lower, even though labor was worked more intensely. Without labor-saving devices and the techniques of scientific management, the planta-tion routine was subject to the same vicissitudes of farming experienced by the smallest homestead: weather, soil conditions, preindustrial tech-nology, and the peculiar problems of working a dependent, antagonis-tic, and seasonally underemployed labor force.

The most that Phillips may claim for his per capita production data is that the emancipated blacks in the Yazoo Delta were less employed in cotton production than had been the case of the delta blacks in slavery. How efficient they were as sharecroppers, fixed rent tenants, or owner-operators is another matter completely. Moreover, the per capita pro-duction of one crop is a nearly useless estimate for measuring the relative, personal efficiency of labor. What else was produced? Finally, because the efficiency of labor is an estimate of how well an individual or group of workers labored with the means of production available, the proper determination of labor efficiency demands some specification of what was used to grow cotton. Was the low per capita production of cotton that Phillips observed in 1913 the result of relatively inefficient labor, or was it the result of a change in the quantity and quality of the other means employed (land, labor, technology, and capital)?

THE NATCHEZ DISTRICT AS A CASE STUDY

This study examines in some detail the origins and institutionalization of sharecropping as it emerged in the American South in the days after slavery. Because the crucial questions to be asked about the emergence and functioning of sharecropping require the comparative analysis of the system to slavery, I have selected the case study approach as the best way of obtaining the necessary details and the proper perspective. No other method is better suited to analysis of the black's comparative efficiency as a sharecropper and slave. But the main thrust of this work is to understand the freedman's motivation in coming to sharecropping.

Obviously, the in-depth study of any locality will not yield conclusions that perfectly reflect the whole. But if the locality is somewhat typical of the whole and, more importantly, relatively self-contained, the case study is a useful and rewarding historical tool. In view of Phillips's caution that the issue of efficiency demands investigating districts where "no great economic change has occurred except the abolition of slavery," what is needed is a locality relatively stable in trade, soil conditions, urbanization, and population in the immediate postwar generation. Several neighborhoods in the late nineteenth-century American South fit this description, but few are more appropriate than Adams County, Mississippi, and Concordia Parish, Louisiana. Besides having used relatively little fertilizer prior to 1890, an important consideration for measuring the relative efficiency of labor, these two political entities formed one historical district that was somewhat characteristic of the large society of which they were a part. Adams County, one of the oldest cotton areas in the South, was not only home to the South's richest plantation elite but was also typical in 1860 of the older South's deep-rooted plantation economy. Directly across the Mississippi River, Concordia Parish, on the other hand, was a newly established area with large plantations, virgin soils, absentee planters, and a relatively isolated slave population somewhat typical of the black belts and delta regions of the Southwest.

The historical importance of the Natchez District makes it worthy of study for that reason alone, aside from its role here as a case study in the origins of sharecropping. In fact, there is a substantial body of literature on the district's antebellum history as well as important work on its twentieth-century economy and society. Most notable is D. Clayton James's *Antebellum Natchez* (Baton Rouge, La.: Louisiana State University Press, 1968) and Allison Davis, Burleigh B. Gardner, and Mary R. Gardner's *Deep South: A Social Anthropological Study of Caste and Class* (Chicago: University of Chicago Press, 1941). The second work is a classic treatment of class and caste in a twentieth-century southern

Plate 1. *The Natchez. Harper's Weekly*, November 12, 1870.

town, Natchez, and ranks among the most important sources for our understanding of sharecropping as a total institution. Yet the transition period between the neighborhood's slave setting and its more recent caste-like society and political economy is one of relative historical ignorance. It is this gap with which this book is concerned.

NOTES

1. See Steven Cheung, *The Theory of Share Tenancy* (Chicago: University of Chicago Press, 1969); and Gunnar Myrdal, *The American Dilemma: The Negro Problem and Modern Democracy*, 2 vols. (New York: Harper & Row, 1944), 1: 230-64.

2. M. B. Hammond, "The Southern Farmer and the Cotton Question," *Political Science Quarterly* 12 (September 1897), pp. 451-65. This approach was characteristic of the writings on the subject prior to 1940. Although Hammond suggested that sharecropping emerged partly because planters lacked the funds for paying weekly or monthly wages, he emphasized in his argument that the freedman's unreliability was the crucial factor involved. In his own words, the share system was adopted in order to force the laborer "to share the losses which are occasioned by his own idleness and neglect." See *Magazine of History* 61 (October 1958), p. 363; Robert P. Brooks, *The Agrarian Revolution in Georgia, 1865-1912* (Madison, Wis.: University of Wisconsin Press, 1914), pp. 1-65; Thomas J. Edwards, "The Tenant System and Some Changes Since Emancipation," American Academy of Political Science, *Annals* 49 (September 1913), pp. 38-46; William C. Harris, *Presidential Reconstruction in Mississippi* (Baton Rouge, La.: Louisiana State University Press, 1967), p. 20; Marjorie Stratford Mendenhall, "The Rise of Southern Tenancy," *Yale Review* 27 (Autumn 1937), pp. 110-29; Fred A. Shannon, *The Farmer's Last Frontier: Agriculture, 1860-1897* (New York: Harper Torchback edition, 1968), pp. 82-88.

3. William S. McFeely, *Yankee Stepfather: A Study of General O. O. Howard and the Freedmen's Bureau* (New Haven, Conn.: Yale University Press, 1968), pp. 149-65. McFeely is the most recent of those historians who emphasize the freedman's hope for land as a significant factor behind the emergence of sharecropping. His work, however, fails to detail the process. See also LaWanda F. Cox, "Agricultural Labor in the United States, 1865-1900, With Special Reference to the South" (Ph.d. dissertation, University of California at Berkeley, 1941); Willard Range, *Century of Georgia Agriculture, 1850-1950* (Athens, Ga.: University of Georgia Press, 1954), pp. 61-81; Joel Williamson, *After Slavery: The Negro in South Carolina During Reconstruction, 1861-1877* (Chapel Hill, N.C.: University of North Carolina Press, 1965), pp. 32-126. Williamson's study is by far the best work to date on the origins of sharecropping.

4. William E. Laird and James R. Rinehart, "Deflation, Agriculture, and Southern Development," *Agricultural History* 33 (April 1968), pp. 115-25; Vernon L. Wharton, *The Negro in Mississippi, 1865-1890* (Chapel Hill, N.C.: University of North Carolina Press, 1947).

5. Stephen J. DeCanio, *Agriculture in the Postbellum South: The Economics of Production and Supply* (Cambridge, Mass.: MIT Press, 1974); Robert Higgs, *Competition and Coercion: Blacks in the American Economy, 1865-1914* (New York: Cambridge University Press, 1977); Robert Higgs, "Did Southern Farmers Discriminate?" *Agricultural History* 46 (April 1972), pp. 325-28; Robert Higgs, "Patterns of Farm Rental in the Georgia Cotton Belt, 1880-1900," *Journal of Economic History* 34 (June 1974), pp. 468-82; Roger L. Ransom and Richard Sutch, *One Kind of Freedom: The Economic Consequences of Emancipation* (Cambridge, England: Cambridge University Press, 1977); Joseph D. Reid, "Sharecropping and Agricultural Uncertainty," *Economic Development and Cultural Change* 24 (April 1976), pp. 549-76; Joseph D. Reid, "Sharecropping as an Understandable Market Response: The Post-Bellum South," *Journal of Economic History* 33 (March 1973), pp. 106-30; Joseph D. Reid, "Sharecropping in History and Theory," *Agricultural History* 49 (April 1975), pp. 426-40; Richard Sutch and Roger Ransom, "The Ex-Slave in the Post-Bellum South: A Study of the Economic Impact of Racism in a Market Environment," *Journal of Economic History* 33 (March 1973), pp. 131-38.

6. Harold D. Woodman, "Sequel to Slavery: The New History Views the Postbellum South," *Journal of Southern History* 53 (November 1977), pp. 524-54.

7. Reid, "Sharecropping as an Understandable Market Response."

8. Higgs, "Patterns of Farm Rental in the Georgia Cotton Belt."

9. See Eugene D. Genovese, *Roll, Jordan, Roll: The World the Slaves Made* (New York: Pantheon Books, 1974); Lewis Cecil Gray, *History of Agriculture in the Southern United States to 1800*, 2 vols. (Washington, D.C.: Carnegie Institute of Washington, 1933), 1: 558-59; Herbert G. Gutman, *The Black Family in Slavery and Freedom, 1750-1925* (New York: Pantheon Books, 1976); Raimondo Luraghi, *The Rise and Fall of the Plantation South* (New York: New Viewpoints, 1978), pp. 64-82; U. B. Phillips, *Life and Labor in the Old South* (Boston: Little Brown & Co., 1929), pp. 186-218; George P. Rawick, *From Sundown to Sunup: The Making of the Black Community* (Westport, Conn.: Greenwood Press, 1972), pp. 53-74; Kenneth Milton Stampp, *The Peculiar Institution: Slavery in the Ante-Bellum South* (New York: Alfred A. Knopf, 1956), pp. 142-91.

10. Mrs. E. P. Pipes to G. R. Reynolds, Provost Marshal, Natchez, Mississippi, April 1, 1965, Records of the Bureau of Refugees, Freedmen, and Abandoned Lands (hereinafter cited as BRFAL), Record Group 105, National Archives, Washington, D.C.

11. Benjamin F. Cherry, Assistant Provist Marshal, Vidalia, Louisiana, to General Lorenzo Thomas, May 5, 1864, BRFAL, Record Group 105.

12. "Relations Between Landlord and White Labor," *The American Farmer* (February 1868), pp. 250-51.

13. See Woodman, "Sequel to Slavery," pp. 551-54.

14. See Herbert G. Gutman, *Work, Culture and Society in Industrializing America* (New York: Vintage Books, 1977), pp. 1-79; E. P. Thompson, *The Making of the English Working Class* (New York: Vintage Books, 1963).

15. See Stanley Elkins, *Slavery* (Chicago: University of Chicago Press, 1958) and Ned O'Gorman, *The Children Are Dying* (New York: Signet, 1978), for examples of two quite different statements of this view.

16. *Intruder in the Dust* (New York: Random House, 1948), p. 11. Faulkner poetically deals with the southern black's "condition" of poverty in this way:

So he stripped off the wet unionsuit too and then he was in the chair again in front of the now bright and swirling fire, enveloped in the quilt like a cocoon, enclosed completely now in that unmistakable odor of Negroes— that which if it were not for something that was going to happen to him within a space of time measurable now in minutes he would have gone to his grave never once pondering speculating if perhaps that smell were really not the odor of a race nor even actually of poverty but perhaps of a condition: an idea: a belief: an acceptance, a passive acceptance by them themselves of the idea that being Negroes they were not supposed to have facilities to wash properly or often or even to wash bathe [sic] often even without the facilities to do it with; that in fact it was a little to be preferred that they did not.

17. Thomas W. Knox, *Camp-Fire and Cotton-Field* (New York: Blelock & Co., 1865), p. 372. Knox claims to have found but few blacks in the Mississippi Valley who imagined that freedom gave them the right to be unemployed or to become vagrants. Although most freedmen expected to labor, "they expected compensation for their labor, and did not look for punishment." See also Jay R. Mandle's *The Roots of Black Poverty: The Southern Plantation Economy after the Civil War* (Durham, N.C.: Duke University Press, 1978), p. 37 for a brief and undocumented argument in support of the interpretation afforded herein.

18. "The Experts," *The Crisis* (March 1913), pp. 239-40.

19. U. B. Phillips, "The Decadence of the Plantation System," American Academy of Political Science, *Annals* 35 (January 1910), pp. 37-56.

20. U. B. Phillips, "The Economics of the Plantation," *South Atlantic Quarterly* 2 (July 1903), pp. 231-36.; "Conservation and Progress in the Cotton Belt," *South Atlantic Quarterly* 3 (January 1904), pp. 1-10. See also C. E.

Allen, "Greater Agricultural Efficiency for the Black Belt of Alabama," American Academy of Political Science, *Annals* 61 (September 1915), pp. 187-98; Enoch Banks, *The Agrarian Revolution in Georgia* (New York: Columbia University Press, 1905), pp. 738-53.

21. U. B. Phillips, "Plantations with Slave Labor and Free," *American Historical Review* 30 (July 1925), pp. 738-53.

22. Gray, *History of Agriculture* 1: 478-80.

23. C. Vann Woodward, *Origins of the New South, 1877-1913* (Baton Rouge, La.: Louisiana State University Press, 1951), pp. 179-80.

24. Robert W. Fogel and Stanley L. Engerman, "The Relative Efficiency of Slavery: a Comparison of Northern and Southern Agriculture in 1860," *Explorations in Entrepreneurial History* 8 (Spring 1971), pp. 364-65; Fogel and Engerman, "Explaining the Relative Efficiency of Slave Agriculture in the Antebellum South," *American Economic Review* (June 1977), pp. 275-94; Jacob Metzer, "Rational Management, Modern Business Practices, and Economies of Scale in the Ante-Bellum Southern Plantations," *Explorations in Economic History* 12 (April 1975), pp. 123-150; Thomas M. Zepp, "On Returns to Scale and Input Substitutability in Slave Agriculture," *Explorations in Economic History* 13 (April 1976), pp. 165-78.

25. See Mark D. Schmitz, "Economies of Scale and Farm Size in the Antebellum Sugar Sector," *Journal of Economic History* 37 (December 1977), pp. 959-80; and especially Harold D. Woodman, *King Cotton and His Retainers: Financing and Marketing the Cotton Crop of the South, 1800-1925* (Lexington, Ky.: University of Kentucky Press, 1968), pp. 14-186.

26. Guy S. Callender, ed., *Economic History of the United States* (New York: A. M. Kelly, 1965), pp. 54-61; U. B. Phillips, *Life and Labor in the Old South* (Boston: Little, Brown & Co., 1929), p. 185. Although a recent interpretation argues that the large slave plantation may have utilized a higher land to labor ratio or better land per hand than did the smaller operations, the costs would have nevertheless increased proportionally. See Fogel and Engerman, "Relative Efficiency of Slavery," pp. 363-64.

27. Fogel and Engerman, "Explaining the Relative Efficiency of Slave Agriculture," pp. 290-94. Metzer, "Rational Management, Modern Business Practices, and Economies of Scale," pp. 123-50

28. Ransom and Sutch, *One Kind of Freedom*, pp. 74-78, argue that it was unlikely that either the use of coercion or the division of labor possible on the large plantation resulted in economies unavailable to smaller farmers. See also Robert R. Russell, "The Effects of Slavery upon Nonslaveholders in the Ante-Bellum South," *Agricultural History* 15 (April 1941), pp. 112-26; Yoram Barzel, "An Economic Analysis of Slavery," *Journal of Law and Economics* (April 1977), pp. 102-104.

29. Metzer, "Rational Management, Modern Business Practices, and Economies of Scale," pp. 123-50.

30. Harry Braverman, *Labor and Monopoly Capital: The Degradation of Work in the Twentieth Century* (New York: Monthly Review Press, 1974), pp. 1-236.

31. Ibid.; Phillips, "Plantations with Slave Labor and Free," pp. 738-53.

32. Gray, *History of Agriculture to 1800*, 1: 479.

33. Ibid. 5 p. 450.

• 2 •

THE ANTEBELLUM SETTING

Before examining the transition from slavery to sharecropping in the Natchez District, we must know something about its antebellum economy in order to analyze those features essential to understanding the changes wrought by the Civil War.

THE LAND

When the pioneer French came to the area in the late seventeenth century, the western half of what became the Natchez District was part of a massive flood plain, a swamp 3 to 25 miles wide, over 50 miles long, and confined on three sides by water. Inundated each year with flood waters from the Black, Tensas, and Mississippi rivers, this marshy wilderness, in what is now Concordia Parish, Louisiana, held little attraction for settlers who much preferred the uplands across the Mississippi River.[1] (See Map 1.) Here, in modern-day Adams County, Mississippi, the French, Spanish, and then British and Americans found bluff lands of yellow clay loam extending into the back country for several days' ride. Except for a narrow section fronting on the Mississippi, few valleys or swamp lands broke the hilly terrain, even though a major river, the Homochito in the south, and four streams drained the uplands. Here was a country with ample means of access, fertile soil, well-drained, and relatively easy to improve and farm—outstanding features in comparison to the marsh lands in the south and west.[2]

But by 1860 the economics of large-scale cotton production with slave labor had had mixed effects on the district's soil. In Adams

LOCATION OF STUDY AREA

0 _____ 100 miles

MISSISSIPPI

LOUISIANA

Adams County

Concordia Parish

Map 1. The Natchez District

25

County, the state's geologist found the land practically beyond production unless manured. Farms once capable of producing 2 to 3 bales of cotton per acre yielded less than half a bale in 1859. Although horizontal plowing was commonly practiced in the county, few planters utilized the technique to constrain erosion but to better work even the steepest hills.[3] As a result, the pleasantly rolling lands of a bygone day had given way to deep gullies and washes by the middle of the nineteenth century.[4]

In other parts of the antebellum South, similar conditions were accompanied by economic hardship and adjustments. This was not the case in Adams County, however, since the once uninhabitable swamplands across the river offered a relatively easy solution to planters in the district's Mississippi neighborhood. A letter written by a reporter on tour of Louisiana in 1859 for the *New Orleans Crescent*, J. W. Door, indicates just how valuable the swamps had become:

> . . . the plantations are magnificent in extent stretching away in broad expanses, level as water, and, at this time, a wavy sea of green cotton plants, variegated with the pink and snowy whiteness of the blossoms and bursting boles; while long ranks of Negroes are working across them gathering in the fleecy harvest of the season.[5]

An intricate system of levees had turned the swampy plain of the delta into rich meadows and cotton fields, with much of the interior similarly drained and in cultivation by 1860. Here, then, was a rich frontier of fertile soils within the very neighborhood of one of the oldest cotton districts in the South. Indeed, only 21 percent of Concordia's land lay improved on the eve of the Civil War, although the slave population in the parish had nearly doubled from 6,934 in 1850 to 11,867 by 1860.[6]

THE PEOPLE

This combination of exhausted lands and old wealth amidst virgin soils and new fortunes, located abreast the county's greatest river, and a multinational heritage had produced one of the South's most colorful neighborhoods. Travelers in the area seldom missed the opportunity of spending a few days in the bustling river town of Natchez, drawn there by its reputation for high living and notorious characters. Frederick L. Olmsted, the Yankee traveler, noted especially the scale and opulence of the district's planter class. Here in a single neighborhood lived and worked the South's exceptionally large planting elite—"swellheads" worth in some cases up to $10 million owning hundreds of slaves and living in grand townhouses and plantation mansions. No other district in the South so impressed this seasoned observer in regard to the wealth of its planting elite.[7]

Historians have also associated the Natchez District with the antebellum elites of southern society. Clayton James found forty aristocratic families in the area, including those like David Hunt and A. V. Davis who owned slaves and plantations scattered over several counties.[8] Morton Rothstein has argued that several of these Natchez District planters acted as full-scale capitalists: shrewd businessmen who efficiently managed their cotton and sugar operations as well as their investments in eastern securities and western lands.[9] In view of such enterprise, Rothstein suggests that the antebellum South is best understood as a dual economy in which its richest planters participated in all aspects of the advanced exchange economy, while its majority white population, as self-sufficient farmers, functioned largely outside of the market.[10]

A close examination of the Natchez District bears out many of these impressions. Seventy percent of the farmers in Concordia Parish and Adams County owned over fifty slaves in 1860, while only 10 percent owned no slaves at all.[11] Moreover, it was not uncommon for planters in the hundred-slave category to own several plantations throughout the lower Mississippi Valley. Absentee owners, mainly Adams County planters, owned 33 percent of the improved lands in Concordia Parish in 1860.[12] A surveyor's map of Adams County (Map 2), drawn sometime in the decade before the Civil War, is vivid testimony to the top-heavy character of landownership in the area. Dozens of large estates, listed simply as Frogpond, Ivanhoe, New Era, or Solitary Valley, occupied almost every acre of an 8-mile radius around Natchez. Plantations grew even larger in size as the circle widened to the east and south, but numerous smaller farms nevertheless competed for space in the interior. Across the Mississippi River in Concordia Parish (Map 3) the same pattern held true. Large plantations such as Whitehall, Minorca, and Sycamore contested the water for survival along the western bank of the Mississippi and around Davis Island and Lake Saint John in the north, while smaller farms cropped the fertile but less accessible back country.

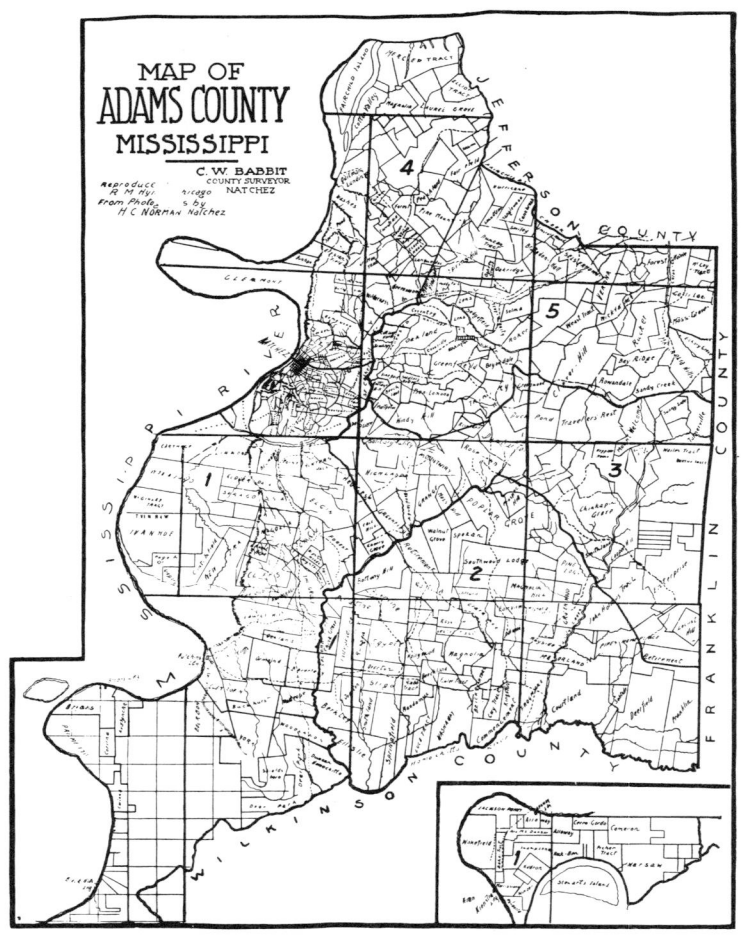

Map 2. Map of Adams County, 1860.

28

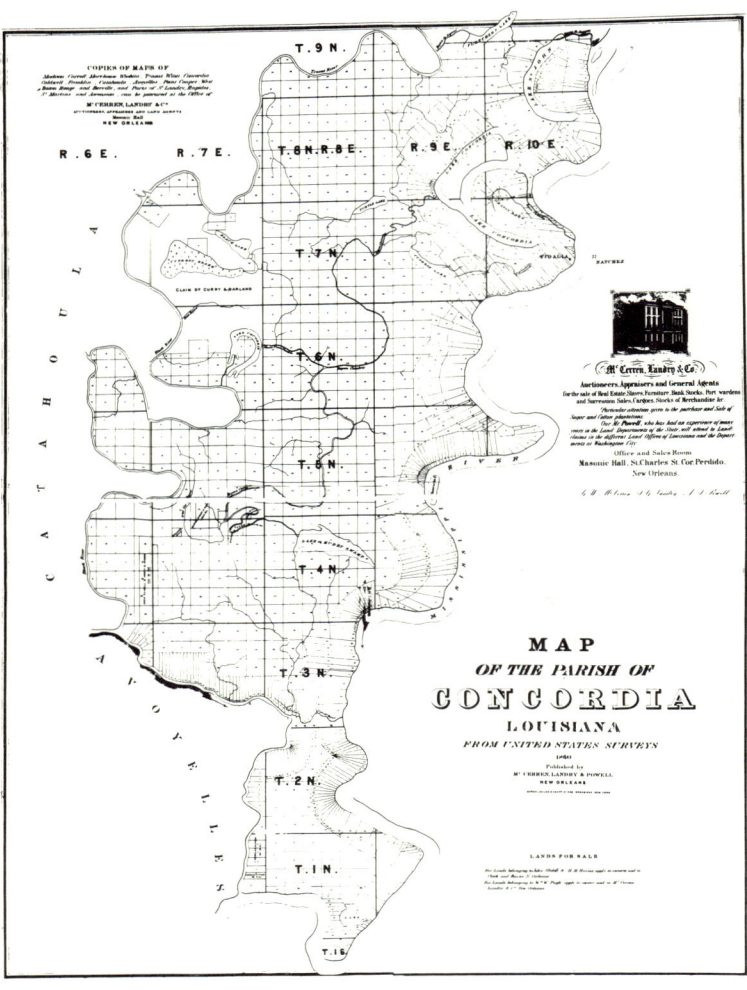

Map 3. Map of Concordia Parish, 1860.

But how typical of their class were those Natchez "Nabobs" who dealt in western lands and northern securities? The answer is largely unknown, since few personal accounts documenting investments have survived. Even the available manuscript tax records generally underestimated local assets while, of course, ignoring outside investments completely. For instance, Stephan Duncan estimated his holdings of northern securities at $479,500 in 1856, but county tax records for 1852 give no indication of Duncan having any investments at all outside of slaves and land.[13] In 1861, for what it is worth, the tax records for Adams County listed seventy-six people with money invested in notes, bonds, bills, or other securities in excess of $1,000. The sums invested ranged up to $58,000, with the ten largest portfolios valued at around $33,000 on the average. Over half the investors held notes worth less than $8,000, but only 12 percent of the county's large slaveowners appeared as investors on the tax sheets.[14]

Although it would be questionable to assume that such investments indicated the propensity of planters to invest outside the cotton economy, they may reveal the extent to which the largest planters acted as bankers or backers in their home area. Here the largest planters were seemingly planters first and foremost, and those with surplus capital to invest apparently put it back into land and slaves rather than into local notes or commerce and trade. With regard to a planter's investments outside the cotton economy, it may be stated that they probably stemmed from the sheer size of planter incomes rather than the bourgeois-acquisitiveness commonly associated with eastern merchant-entrepreneurs. David Hunt, for instance, inherited in the 1830s a chain of stores and gins stretching from New Orleans to Vicksburg which he eventually converted into twenty-seven plantations, seventeen hundred slaves, and extensive investments in western lands and railroad securities.[15] The point, of course, is whether the transition from trade to planting on such a large scale involved preferences unrelated to sound business sense. The scope of such plantation enterprise suggests that Hunt may have invested outside the plantation economy after having literally exhausted the opportunities available on the local scene. (Twenty-seven plantations undoubtedly taxed his managerial abilities to the limit.) In this sense, his outside investments may have been a function of his surplus capital rather than his seeking the highest returns for his capital. That other planters also enjoyed above-average incomes and yet apparently avoided participating in some aspects of the economy (local banking or commerce and trade) while acting in others (western lands, eastern securities, and the slave plantation economy) denotes something other than the process suggested by Rothstein. Natchez planters conceivably shared a common set of attitudes about investments and enterprise that are not best described in terms of men shrewdly participating in all aspects of a mature exchange economy.[16]

This nonbourgeois character of the area's plantation elite has been criticized by historian Clayton James who points to a more than coincidental relationship between the concentration of Nabobs in Natchez and the district's failure to develop as a trade center. In the mind of one contemporary observer, their wealth had even spoiled the Nabobs as slave masters. F. Lloyd King, a Georgia antebellum planter who leased land in the district after the Civil War, expressed shock at the "purse proud, ill-educated, raw, drinking, gambling bravos" of Natchez, "who, when they were rich, spent their money in Negroes and in dissipation and left their debts unpaid." According to King, slaves in the Natchez District were bought and sold for reasons of speculation with little concern for their welfare. At home on the coast of Georgia, slaves were considered part of an estate: "We neither bought nor sold, that is nothing of the kind was done on our plantation during my life."[17]

The truth of the matter is more complex than these impressions. Some planters, like Duncan, undoubtedly functioned as full-scale capitalists interested in the pursuit of profits regardless of their scope and setting. He loaned money to southern planters if he felt they were good risks, joined in partnership with local bankers, and carefully instructed his eastern brokers as to the handling of his finances. Other planters, like those described by King, gave the slave mart in Vidalia, across the river from Natchez, a brisk business and probably traded in the human chattel with an eye to speculation. Perhaps Duncan did the same. But most Natchez Nabobs were primarily farmers in the business of cotton planting with slave labor. The largest of them grew rich, invested their profits in slaves, and lived conspicuously in grand style. As absentee owners of several plantations, they undoubtedly bought and sold slaves with an abandon unmatched by their smaller neighbors. Yet, most of them were also resident planters with at least one home place as the basis of their business, life, and society. Here they may have more closely matched what King felt was the norm in Georgia: planters whose antebellum enterprise possibly involved conceptual limitations best understood by viewing them as slave masters first and foremost.[18]

Although few travelers bothered to comment on the district's lesser folk, men who owned twenty to fifty slaves farmed throughout the neighborhood. Interestingly enough, Concordia Parish contained fewer small planters than the more well-established county across the river, and those in Louisiana tended to be older as a group.[19] Common sense might lead one to expect older age patterns in the more established Adams County, especially since the majority of small planters east of the river came from Mississippi, unlike those in Concordia where 70 percent hailed from neither Louisiana nor Mississippi. The pattern denotes, of course, the practice among Adams County planters of dividing portions of their estates among sons and heirs. Such was clearly the case with James and Robert Pipes, young men in their twenties, who were set up by their wealthy planter father, Lewis Pipes, with two dozen slaves each and enough land to get a start. What is more, the large number of small planters living in Adams County is misleading in that a fifty-slave plantation might be owned by someone with larger holdings elsewhere, but living on a relatively small family estate close to Natchez. The typical forty-slave planter in Concordia, on the other hand, coming from outside the state, seldom owned property other than the place he farmed.

The next group, farmers owning one to twenty slaves, varied in character from the above pattern. In this cluster, the Adams County farmers included enough elderly to make them older as a group in comparison to those in Concordia, and a large number, almost 50 percent, had migrated from parts other than Mississippi or Louisiana. Concordia farmers were younger, however, in comparison both to those one step up the ladder in Louisiana and to those of similar status in Adams County. In addition, 90 percent of the Concordians in this cluster came from either Louisiana or Mississippi, with many having inherited land as second-generation farmers in the area. But it is doubtful that the heirs in this group received much beyond small family plots in the back country around the Tensas River. That so many of the Adams County group, on the other hand, were first-generation farmers indicates that more than a few Mississippi planters had inherited enough by 1860 to place them well into the middle planter category.

The final set of farmers reported in the agricultural census, those without slaves, were the oldest of the groups in Adams County, forty-six years, and the second oldest in Concordia, forty-five years. Relatively few of these men worked solely as farmers, and few hailed from either Louisiana or Mississippi. Self-declared blacksmiths, laborers, carpenters, and stock raisers, they nevertheless farmed enough to have their produce listed in the census.[20]

Included, too, among the population were the nonlandowners: overseers, slaves, and merchants. The overseer, who usually owned neither slaves nor land, played a key role in the district's plantation economy, especially in Concordia Parish where the entire system depended on his ability to care for the absentee-owned estates.[21] As with the farmers, several important differences appear in respect to the overseer class when Concordia Parish is compared to Adams County. First, almost three-fourths of the farms on the Louisiana side employed overseers in contrast to the one-fourth across the river in Mississippi, reflecting, of course, the absentee nature of landownership in the parish. Second, a much higher percentage of Concordia overseers owned substantial real estate and personal property than did their counterparts in Adams County. Indeed, the profile in Table 1 suggests that Mississippi overseers were far below the regional average in this aspect, and in terms of property and slaves owned even the most successful fade in contrast to those in Concordia. William Shields, for instance, the long-time overseer of the three Mercer plantations in Adams County, owned only $3,000 in personal property and no real estate or slaves on the eve of the Civil War.[22] The reasons for this difference seem clear enough: working in what amounted to a frontier area, overseers in Concordia had some opportunity to buy lands and work their own slaves while still managing the estates of others. The fact, however, that fewer than one-fourth of the Concordia overseers owned land or slaves in 1860 indicates the limitations experienced by this class in general.

Although the much noted tendency to discord between overseer and planter undoubtedly existed in the Natchez District,[23] it hardly affected the plantation's productivity. The largest farms in Concordia Parish—those most apt to employ overseers—were larger by any measure used and more productive than those in Adams County. While the difference is largely explained by Louisiana's relatively fertile soil, it is evident that absentee owners employed the overseer class to great success. The overseer's role as manager enabled the wealthiest planters in the district to coordinate several operations in a single enterprise. The case of J. A.

Table 1
Overseers by Age and Property in Adams County and Concordia Parish
in 1860 as Compared with Composite Regional Averages[a]

Overseer Attributes	Concordia, La.: 137 Overseers	Adams, Miss.: 50 Overseers	Composite Regional Average: 808 Overseers
Average Age	34.2	33.9	30.6
Percent over 40 years	28	30	14
Percent owning personal property	86	14	44
Percent owning substantial personal property	11	4	8
Average age of those owning substantial personal property	37.3	43.5	35.3
Percent owning real property	17	4	10
Percent owning substantial real property	14.5	4	8
Average age of those owning substantial real property	38.5	27	34.2
Percent born outside of Louisiana or Mississippi	62	48	

[a] The regional averages are taken from William Kauffman Scarborough, *The Overseer: Plantation Management in the Old South* (Baton Rouge, La.: Louisiana State University Press, 1966).
SOURCES: Manuscript Census; Scarborough, *The Overseer*, pp. 62-63.

Gillespie is a telling one (see Table 2). The old planter owned a plantation in Adams County, Hollywood, where he lived, and another in Concordia, Indian Village, managed by a trusted overseer. His Louisiana property, although less productive in corn, far exceeded his home plantation in cotton production. Hollywood served as home and hearth while his Indian Village functioned as the workhorse of Gillespie's enterprise in terms of cotton and income. In addition, Gillespie's Mississippi plantation may have produced more corn than it needed, with the surplus going to his Concordia place, thus enabling the latter to better concentrate on a cash crop.[24] Thus, much was expected of the overseer if the two units were to complement one another efficiently.

Table 2
Average Production of Cotton and Corn, Average Holdings of Improved Acres on the Largest Plantations in Adams County and Concordia Parish in 1860, and on the Two Plantations Owned by J. A. Gillespie

			Gillespie Plantations	
			Indian	
	Adams, Miss.:	Concordia, La.:	Village	Hollywood
	70 Plantations	90 Plantations	(Concordia)	(Adams)
Average number of slaves per plantation	103	112	100	91
Average bales of cotton per plantation	180	612	700	125
Average bushels of corn per plantation	3,341	4,298	3,000	4,000
Average number of improved acres per plantation	777	896	800	600
Average bales of cotton per slave	2.6	5.4	7.0	1.3
Average bushels of corn per slave	32	38	4	32
Average number of improved acres per slave	7.4	7.9	8.0	6.5

SOURCE: Manuscript Census.

Good and Faithful Labor

The most indispensable group in the area's agricultural population plowed the fields, planted the seed, picked the crop, and bore the economy's heaviest burdens. In 1860, slaves outnumbered the white population eight to one in Concordia Parish and four to one in Adams County.[25] Most of them worked on large plantations in a life of unyielding toil under some of the harshest conditions in the South. The work in Concordia involved a never-ending battle with disease, snakebite, heat, and the surrounding swamp waters. At least in the uplands, once a field was cleared it stayed that way, and few planters broke much new ground in the county in the decade before the Civil War. But Concordia was largely a virgin wilderness in the 1850s. Slaves and masters alike worked to clear the lands, to keep the levees intact, and to stop the underground seepage that could spread over freshly planted fields in a twelve-hour period.

Slaves in the two parts of the Natchez District probably experienced a similar way of life; yet, as shown in Tables 3 and 4, there were significant differences. Fifty-five percent of the slaves in Concordia Parish lived on large plantations in comparison to 38 percent in Adams County. Many more of the Louisiana places had absentee owners and were worked by overseers. Except for the house servants, of which there were fewer in Concordia, the typical field hand on the swamp plantations perhaps only vaguely knew his owner or the owner his slave. For the Adams County slave, there were more whites to contend with, closer supervision, and a more complex society on the whole. Natchez was accessible, and so, too, was the main highway or Natchez Trace which brought travelers of every sort and character through the neighborhood. A larger percentage of Adams County slaves worked as house servants with relatively high personal skills and some refinement, as is to be expected of those who retain the fashionable and rich. Adams County slaves were responsible for the care of plantation mansions and townhouses, which they themselves often built. Hence, they experienced a diversity in their lives seldom known by the swamp chattel.[26]

The isolation of the large swamp plantations, however, may have fostered a more viable and more deeply rooted slave community. Historian C. Peter Ripley has found strong evidence of stable family relations in his examination of slave marriages performed in Union-occupied Concordia Parish in late 1864 and early 1865. A large number of Concordia freedmen wanted their slave family ties officially recognized,

Table 3

Number and Percent of Farms, Slaves, and Acres Improved, by Size of Farm or Plantation as Measured in Slaves, Adams County and Concordia Parish, 1860[a]

Category of Measurement		Number of Slaves per Farm or Plantation										
		0	Pct.	1-50	Pct.	51-100	Pct.	101-200	Pct.	210+	Pct.	Total
Number of farms	A	10	4.8	123	59.4	50	24.1	20	9.6	4	1.9	207
	C	23	11.2	91	44.3	46	22.4	40	19.5	5	2.4	205
Number of slaves	A			2,754	26.3	3,615	34.6	2,796	26.7	1,281	12.2	10,446
	C			1,714	14.4	3,447	29.0	5,528	46.6	1,172	9.8	11,861
Acres Improved	A	1,272	1.3	33,110	36.0	30,214	32.8	18,820	20.4	8,500	9.2	91,916
	C	622	0.6	12,745	13.5	32,464	34.5	40,905	43.5	7,290	7.7	94,026

[a]Percentages have been rounded off to the nearest tenth of 1 percent.
SOURCE: Manuscript Census.

Table 4
Statistical Profile of the Slave Population in Natchez District, 1860

Characteristics of Slave Population	Adams (9,241)[a] Pct.	Concordia (10,510)[b] Pct.
Prime hands, 16-65	77.0	69.5
Children, 0-15	22.4	28.3
Old, over 65	0.6	2.2
Female, adults, 16-65	37.5	32.6
Male, adults, 16-65	39.5	36.9
Children on farms of 1-20 slaves	29.4	25.6
Old	2.0	2.6
Female, adults	35.6	26.1
Male, adults	33.0	45.7
Children on farms of 21-50 slaves	23.4	27.2
Old	1.6	0.6
Female, adults	38.3	34.0
Male, adults	36.7	38.2
Children on farms of 51-100 slaves	19.9	27.8
Old	1.6	2.2
Female, adults	36.2	33.9
Male, adults	42.3	36.1
Children on farms of over 100 slaves	19.7	36.4
Old	1.8	3.0
Female, adults	40.2	28.8
Male, adults	38.3	31.8

[a] This figure is 88 percent of the total slave population in Adams County in 1860.
[b] This figure is 89 percent of the total slave population in Concordia Parish in 1860.
SOURCE: Manuscript Census.

and of those previously separated from spouses only 35.7 percent indicated that the "divorce" had resulted from force.[27] We have no way of knowing how this compares to the durability of marriages among slaves in Adams County, but the fact that the largest Concordia plantations listed a far greater percentage of children (36.4 percent to 19.7 percent) is some indication of the relative strength of the slave family in Concordia, unless we assume slave breeding practices unrelated to the family structure. The differences in the number of children follow the above pattern until the one- to twenty-slave unit is reached, although not so great a disparity is noted below the very largest places.

The U.S. Army's experience in employing freedmen as wage hands on leased and abandoned plantations during the war also supports the notion of a relatively rooted community in the swamps. During the war freedmen in Concordia, unlike those across the river, often stayed on the plantations even after the overseers had fled, although they usually resisted being taken away to Texas by hiding in the bush. They stayed near their plantation homes, awaited the arrival of Union troops, reestablished themselves on their home places to live as best they could, or joined together in groups in common defense and in working abandoned plantations under lease from the U.S. Army. More will be said about this subject later, but suffice it here to point out that the superintendent of freedmen in the district noted that the "little community commonly known as fellow servants" on swamp places made it easier for the army to find labor for the Concordia plantations than for Adams County.[28]

Within the Natchez District, the town of Natchez served its planters and farmers as their chief depot and market, although smaller towns were sprinkled throughout the interior. A working population of 1,523 adult whites and 2,229 blacks of all ages serviced gambling houses, saloons, and the river and lumber industry on the docks below the bluff at the Natchez landing, as well as the fine hotels, shops, stores, and mills overhead in Natchez proper. This thriving workaday Natchez with its boatmen and draymen and craftsmen posed a dramatic contrast to the genteel ways of its planter class. The sight of ·fine carriages conveying ladies and gentlemen from their "Stanton Halls" down to "cotton square" to "shop" for mahogany cabinets, yards of calico, and silken sheets, or possibly to the ferry for a visit to the slave mart in Vidalia, provided a spectacle equal indeed to the goings-on below the bluffs in the notorious "Natchez-under-the-hill." One group of men, the merchants, was at home in both worlds.

Eighty-eight storemen, or 77 percent of all storemen in the neighborhood, had stores in Natchez.[29] Although the largest of them dealt directly with the wealthiest planters in the area, few storemen superseded the role of coastal factors but functioned rather as their agents or dealt in

Plate 2. Hiding in the Louisiana Swamps. From *An Album of Reconstruction* by William Loren Katz, copyright 1974. Used by permission of Franklin Watts, Inc.

Plate 3. Natchez Under-the-Hill.

41

cotton on their own accounts for resale to a New Orleans house. Typically, a New Orleans factor provided his planter customer with a line of credit through a local store or commission merchant in Natchez. In return for the business, Natchez merchants collected the cotton consigned for shipment, filled orders over the counter and in semiannual lots, and billed the factor. The number of stores handling such business, combined with the general understanding that factors of repute normally stood behind the planter's debts, enabled the farmer of size to shop around and buy accordingly.[30]

Besides servicing planters, the larger storemen competed with smaller merchants for crops that were too small to be directly consigned to a New Orleans house and for the cash purchases of wealthy planters who needed furniture, apparel, groceries, and books. The large merchants also functioned as wholesalers of goods to country storemen and backcountry peddlers, receiving cotton in return, which they then consigned to their coastal factors. (See Table 5 for data on the number of merchants and the value of merchandise in the decade before the Civil War.)

The far-flung geographic area available to Natchez merchants who wanted to invest in cotton planting directly through landownership and slaves makes it nearly impossible to estimate their involvement or worth. Although careful study of the manuscript census indicates that few merchants in the district owned many slaves in 1860, twenty-one owned personal property worth more than $20,000—a sufficient basis for investment in land and slaves elsewhere. Frederick Stanton, for instance, combined both ventures into one successful career, but his case was atypical to say the least. At the time of his death in 1859, Stanton, while associated with his brother in the cotton commission business, owned 444 slaves and six plantations.[31]

But the majority of Natchez merchants owned too few assets to be considered possible members of the district's plantation elite. Few of these men—those with less than $20,000 in personal property— were involved in much beyond their trade with small farmers in the area. Only one, for instance, as shown in the tax records, loaned money at interest. That they managed to compete at all with the larger merchants is testimony to their drive and the town's extensive hinterland.[32]

Table 5
Number of Merchants, Value of Merchandise of Sold, and Persistence of Merchants in the Natchez District, 1852 and 1861

Value of Merchandise Sold by Date	Number of Merchants	Value of Merchandise	Percent of Total Value	Percent of Total Merchants	Percent in 1852 Remaining in 1861
Under $ 1,000					
1852	31	$ 17,538	1.1	29.0	16.1
1861	37	22,342	2.1	35.6	
$ 1,001-$5,000					
1852	29	$ 76,234	5.0	27.2	27.5
1861	26	80,098	7.4	25.0	
$5,001-$10,000					
1852	13	$ 108,716	7.1	12.1	38.0
1861	12	105,289	9.8	11.5	
$10,001-$25,000					
1852	13	$ 245,080	15.9	12.1	46.1
1861	17	298,788	27.8	16.3	
$25,001 and over					
1852	21	$1,088,824	70.9	19.6	61.9
1861	12	569,700	52.9	11.5	
Total merchandise sold					
1852	107	$1,536,392			34.5
1861	104	1,076,214			

Source: Manuscript Tax Rolls (1818-1861), Adams County, Mississippi.

SELF-SUFFICIENCY AND ECONOMIES OF SCALE

The fact that Natchez merchants advertised their stocks of "500 sack-bushels of Missouri Corn" in local newspapers leads the historian to wonder about the self-sufficiency of the district's larger plantations.[33] To judge from the newspapers, Natchez merchants did a brisk business selling western foodstuffs to district planters. Yet, most Nabobs valued high corn yields in their production plans. One might assume that locational advantages enabled district planters to buy western supplies at less cost than interior farmers, but it might be asked to what extent such advantages undermined the plantation's self-sufficiency in grain needs.

Another question centers on what factors determined the ultimate size of plantations in the Natchez District. As seen in Table 6, the vast majority of district slaves lived and worked on plantations in the above fifty- but below two hundred-slave category, but why this particular distribution? High incomes obviously provided sufficient incentive for expansion beyond the small plantation, but it is uncertain whether farms in the above range experienced efficiencies or economies unavailable to smaller farms or the very large plantations.

Table 6
Distribution of Slaves by Farm Size, Adams County and
Concordia Parish, 1860

Farm Size (in number of slaves per farm)	Adams Pct.	Concordia Pct.
50–100	33.6	28.0
100–200	26.7	46.5
201+	11.3	9.8
TOTAL	71.6	84.3

SOURCE: Manuscript Census.

One way of measuring the large plantation's self-sufficiency is to estimate consumption needs in grains, specifically corn, and to compare these with the census reports of grain production on the individual farms. Such an estimate, however, is crude, and the results uncertain. We know very little about the actual diets of southern people and beasts, and the attempt to be very specific as to grain needs confronts the historian with the difficulties of measuring things that were never enumerated, such as the contribution of fish to the diet of slaves in river neighborhoods or the amount of grain consumed by swine and chickens and rats. But the growing body of literature dealing with the subject includes some consensus about the upper and lower limits of grain needs.[34]

A glance at the data in Table 7 suggests that the larger plantations in Concordia nearly met their grain demands, while the same-size operations east of the river fell seriously below their wants. (See also Appendix A, table 1). Even if the needs have been overestimated by 10 percent (a likely possibility, since the figures neither include peas and beans nor take into account the extent to which animals foraged for themselves in gullies or on the nonenumerated pumpkins and cornstalks), it seems clear that the largest Concordia plantations generally achieved self-sufficiency in grain. More importantly, there is little here to support the theory that absentee owners generally emphasized foodstuffs on their Mississippi plantations while concentrating on cotton in Concordia. In fact, the Concordia plantations apparently produced only enough for their own needs, although it is safe to assume that some may have produced a surplus for use on resident farms across the river in Adams County. There is no evidence that small farmers in the district produced enough to supply both themselves and large plantations in the area.

But who, then, bought those "500-sack bushels of Missouri Corn"? There are many possibilities since Natchez merchants included in their accounts farmers and country stores from a radius reaching far beyond the district. In the neighborhood itself, however, if the estimates are correct, almost all Adams County farmers needed to supplement their grain harvests in 1860. Furthermore, since the data in Table 7 average together individual farmers, it should be obvious that even many large Concordia planters sometimes purchased grains to tide them over until harvest or as a result of short crops. In a word, Natchez merchants sold corn to all groups because of the district's overall grain deficit, but the general self-sufficiency of the larger Louisiana plantations undoubtedly limited their market.

With regard to the possible economies or efficiencies in relation to plantation size, figure 1 leaves little doubt that cotton production per hand increased as farms grew in size beyond the thirty-slave planta-

Table 7
Estimated Grain Needs (Corn) Compared to Grain Production, Natchez District, 1860

Farm Size in Slaves	Area	Corn Grain Needs	Actual Corn Produced	Production of Corn in Relation to Needs (Pct.)
0	Adams	7,974	5,127	65
	Concordia	20,887	2,202	10
1–20	Adams	68,220	50,758	74
	Concordia	57,596	37,462	65
21–50	Adams	123,242	80,480	65
	Concordia	62,930	42,510	67
51–100	Adams	183,656	147,702	80
	Concordia	167,841	137,300	82
100+	Adams	174,545	113,045	65
	Concordia	261,309	260,543	98
Total	Adams	627,118	397,413	63
	Concordia	570,630	480,280	84

SOURCE: See Appendix A.

46

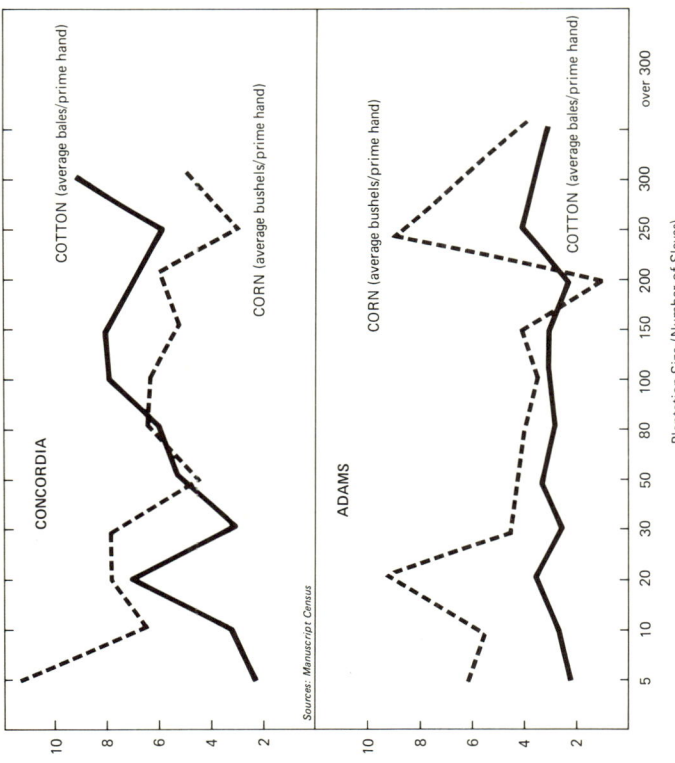

SOURCE: Manuscript Census.
See Appendix B.

Figure 1. Average Cotton and Corn Production, Natchez District, 1860.

tions, only to fall off beyond the one hundred-fifty-slave plantation. These middle-size plantations may have achieved economies of scale, but to be certain we must examine the degree to which soil conditions, crop specialization, and investments in tools, animals, and labor affected production patterns in the Natchez District.

Essentially, the problem is to measure shifts in production, or output, in relation to shifts in costs, or input. If productivity increased while costs remained stable or increased at a less rapid rate or actually declined, then savings may have accrued as a result of size itself, that is, because of better management or the introduction of labor-saving devices which were possible only on the larger farms. Since planters of every size in the Natchez District emphasized cotton and corn almost to the neglect of all other crops, it is a comparatively easy task to measure production shifts relative to farm size by employing the production data found in the manuscript census of 1860.

Although the procedure simply requires dividing total yields and means (computed categorically) for the relevant size groups (defined in terms of slaves) by the number of prime hands (defined here as adults, ages ten to sixty-five), several difficulties are involved. In the first place, by measuring production and costs in terms of prime hands, we may unwittingly bias the results, since shifts in production might have resulted from the number of child laborers on that group of farms. A farm with a relatively small number of prime hands, but nevertheless one working many children in the fields at very young ages, might thus have a higher output per prime hand than one with a relatively larger number of prime hands but few or no children working in the fields. Nor is it sufficient to assume that all children below ten years of age labored to little consequence, since some undoubtedly did while others did not. Accordingly, although the evidence available indicates that children contributed little essential labor, it is nevertheless necessary to compute the precentage of prime hands in the labor force as an input variable.

Second, because the value of the land farmed appears in the manuscript census as the total worth of improved and unimproved acres, it is difficult to determine whether or not changes in labor productivity resulted from changes in the quality of the soil farmed. To circumvent this problem somewhat, I have relied on Gavin Wright's methodology for estimating the value of improved acres, which in Wright's opinion is the best indicator of soil quality.[35] Wright found that by assigning somewhat arbitrary but generally substantiated values to unimproved acreage, and then subtracting that amount from the total farm value listed, a reliable measure of soil quality could be had. The device is clearly susceptible to error in the sense that the values assigned to the unimproved acres may be false. But as the values found in Wright's study generally conform to the values occasionally listed for unimproved land sales in the Natchez District, in my calculations I have employed Wright's average value of unimproved lands, based upon his regional sample, of $3 an acre for Concordia Parish (unimproved alluvial).

If, then, we look at Concordia Parish in 1860 (see figure 2), it seems that planters who owned between thirty-one and eighty slaves, or 22 percent of the total farms observed in the parish, may have achieved some economies because of size. In almost every other group, the variation in cotton production per hand is partly explained by a counter variation in corn production. A similar pattern appears in the amounts of money invested in tools, animals, and land. Farmers in the thirty-one to eighty-slave category generally invested less without farming significantly better soils while obtaining higher yields in comparison to farmers in the other categories. But it should be noted that Figure 4 indicates that this group of planters held somewhat larger proportions of children in their work force relative to those farmers who owned twenty-one to thirty slaves, thus possibly inflating the estimates of their productivity.

In Adams County (see figure 3), farmers who owned thirty to one hundred and fifty slaves—60 percent of the total observed—invested less in tools and work animals as they expanded their enterprise while managing, nevertheless, to hold their cotton production stable. Nor did they experience drastic reductions in corn or much variation in the value of soils farmed. In addition, the percentage of children in the labor force actually declined (see figure 4), thus reducing the bias effects of this factor on the data.

The evidence that some planters may have enjoyed economies of scale is not to say that these savings were anything but accidental. As mentioned earlier, the probability that more than a few planters in the Mississippi half of the neighborhood had inherited estates suggests that plantation size was often unrelated to the planter's appreciation of scale economies. In Concordia, planters generally bought huge chunks of land, began work immediately with a slave force that neither grew nor declined much in size through the years, and either ignored or failed to

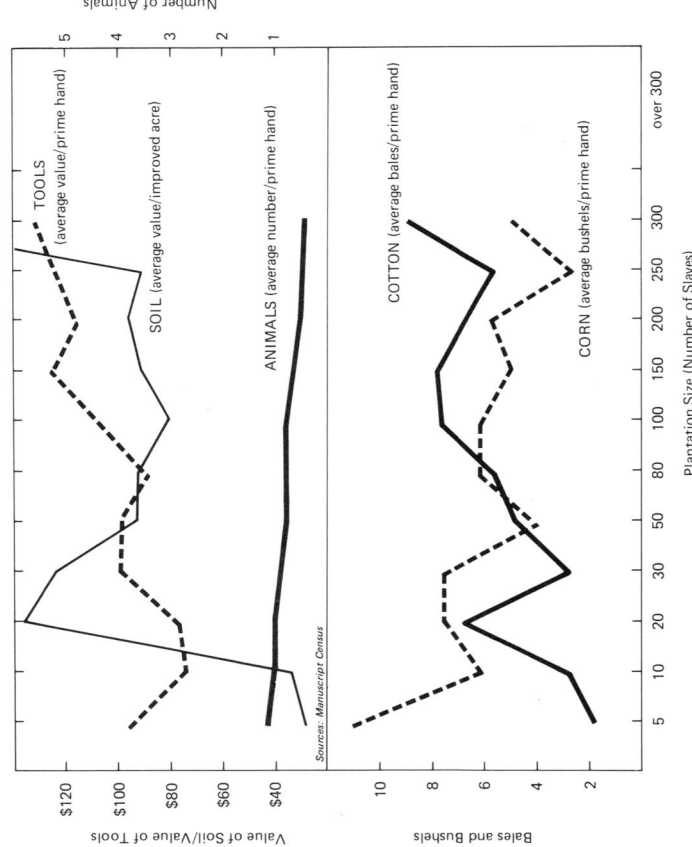

SOURCE: Manuscript Census.
See Appendix B.

Figure 2. Input-Output Data, Concordia Parish, 1860.

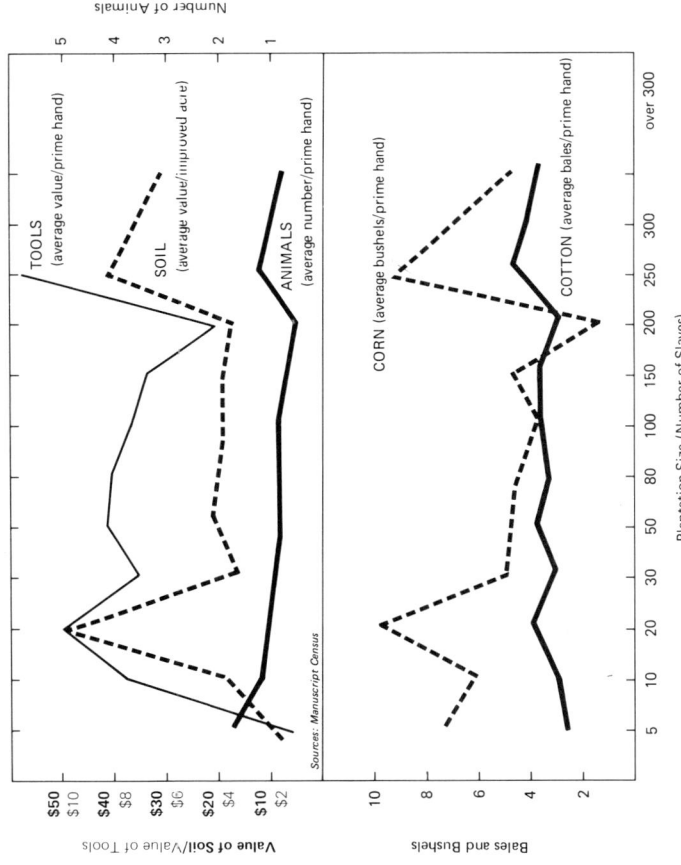

SOURCE: Manuscript Census.
See Appendix B.

Figure 3. Input-Output Data, Adams County, 1860.

51

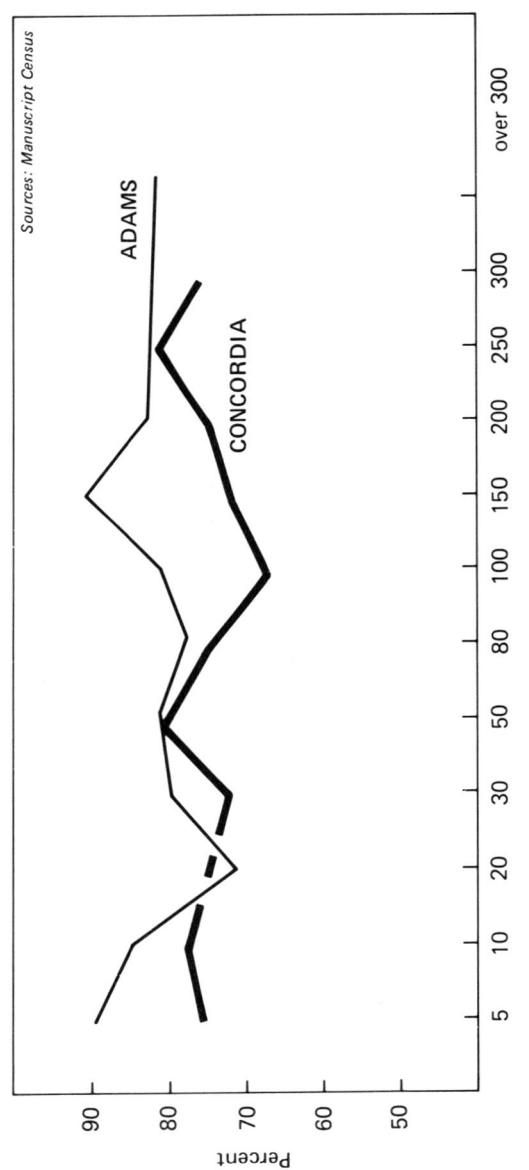

Sources: Manuscript Census

SOURCE: Manuscript Census.
See Appendix B.

Figure 4. Prime Hands (ages 10-65) as Percent of Labor Force.

notice the diseconomies associated with working a force beyond eighty in number. Those aware of the increasing costs per unit of output undoubtedly understood that the expense could be well afforded in view of the higher incomes and status derived from owning and working large numbers of slaves.

On the other hand, even the largest planters limited their slave population to about one hundred and fifty as a maximum number on each plantation. But a plantation's ultimate size reflected complex decisions and varied circumstances. The fact that A. V. Davis, one of the largest planters in the district, held from 104 to 228 slaves on each of his four plantations is indicative of the problems in understanding the motivation behind the size of a plantation's slave force. Why would this prosperous resident of Natchez so distribute his slaves when production information available to himself, the census takers, and future historians suggested that his Potowament place produced more for less in comparison to the other three? Did he even view the situation in terms of separate and competing units of operation? Perhaps he used Sycamore as the central unit for all ginning and final handling of the harvest before shipment. This would explain the larger labor force but not the smaller value of implements and machinery found on Sycamore (Table 8). It is more likely that he concentrated the largest and best of his labor force—defined here in terms of prime hands—on Sycamore in order to take advantage of its immensely fertile soils. There is also the possibility that Davis may have selected the distribution he did primarily with an eye to his heirs. All we know for certain is that the old patriarch, while no ordinary planter, was hardly unique in terms of the size, organization, or distribution of his labor force.

CONCLUSION

In summary, then, here in a single neighborhood was a veritable replica of the South's antebellum plantation economy: large, medium, and small plantations functioning at various levels of self-sufficiency and economy; exhausted soils and fertile new bottom lands; absentee owners and a large resident planter class; thousands of slaves living and working in a diverse environment, ranging from Natchez to relatively isolated community plantations; an extensive class of overseers and managers; merchants of every type and size; and a substantial number of

Table 8

Output-Input Statistics Per Hand on the A. V. Davis Plantations, 1860

Measures of Output-Input	Sahara: 107 Slaves	Potowament: 178 Slaves	Sycamore: 228 Slaves	Tacony: 124 Slaves
Cotton per prime hand (bales)	7.3	8.3	6.4	6.7
Corn per prime hand (bushels)	68.3	48.7	21.5	21.9
Animals per prime hand	0.7	0.6	0.6	0.8
Value of tools per prime hand (cash value)	$105.94	$38.02	$34.94	$89.03
Value of land per prime hand (cash value)	$111.28	$70.05	$145.76	$95.16
Percentage of children (0-10 years)	32.7	32.0	18.4	25.2
Percentage of prime hands (10-65 years)	67.3	68.0	81.6	74.8

SOURCE: Manuscript Census.

small farmers. In such a neighborhood, the historian has available an especially appropriate state for viewing at close range the transition from slavery to sharecropping in the post-Civil War era.

NOTES

1. Robert Dabney Calhoun, "A History of Concordia Parish, Louisiana," *Louisiana History Quarterly* 15 (January 1932), pp. 44-67; 15 (April 1932), pp. 214-33; 15 (July 1932), pp. 428-52; 15 (October 1932), pp. 618-45; 16 (January 1933), pp. 92-124.

2. B.L.C. Wailes, *Report on the Agriculture and Geology of Mississippi Embracing a Sketch of the Social and Natural History of the State* (Mississippi: E. Barksdale, State Printer, 1854), pp. 1-123.

3. E. Hilgard, *Report on the Geology and Agriculture of the State of Mississippi* (Mississippi: E. Barksdale, State Printer, 1860), pp. 315-27.

4. L. Harper, *Report on the Geology and Agriculture of the State of Mississippi* (Mississippi: E. Barksdale, State Printer, 1857), pp. 1-33; John Hebron Moore, *Agriculture in Ante-bellum Mississippi* (New York: Bookman Associates, 1958), pp. 25-38.

5. Quoted in Lewis Atherton, *The Southern Country Store* (New York: Greenwood Press, 1968), p. 19.

6. U.S. Bureau of the Census, *Eighth Census of the United States: 1860. Agriculture*, 3: 255.

7. Frederick Law Olmsted, *The Cotton Kingdom: A Traveller's Observations on Cotton and Slavery in the American Slave States* (New York: Mason Brothers, 1861), pp. 405-36.

8. D. Clayton James, *Antebellum Natchez* (Baton Rouge, La.: Louisiana State University Press, 1968), p. 136.

9. Morton Rothstein, "The Antebellum South as a Dual Economy: A Tentative Hypothesis," *Agricultural History* 41 (October 1967), pp. 373-82.

10. See Harold Woodman's discussion of Rothstein's hypothesis in his "The State of Agricultural History" now published in Herbert J. Bass, ed., *The State of American History* (Chicago: Quadrangle Books, 1970), pp. 65-66.

11. U.S. Census (1860), Manuscript Population and Slave Schedules, Adams County, Mississippi, and Concordia Parish, Louisiana.

12. Herbert Weaver, *Mississippi Farmers* (Nashville, Tenn.: Vanderbilt Press, 1945), p. 111.

13. See Paul W. Gates, "Southern Investments in Northern Lands Before the Civil War," *Journal of Southern History* 5 (May 1939), pp. 155-85; Manuscript Tax Rolls (1818-1861), Adams County, Mississippi.

14. Manuscript Tax Rolls (1818-1861), Adams County, Mississippi.

15. James, *Antebellum Natchez*, pp. 157-58.

16. This point is somewhat supported in the recent study of Davis Bend plantations by Janet Sharp Hermann. She tells the story of how Joseph Davis left a successful legal career in Natchez to become a rural planter. His choice was typical in the antebellum South in that becoming a planter was all that really mattered. Historians T. H. Breen and Aubrey C. Land have shown that southern merchants made similar decisions from the very beginnings of the plantation experience. Breen finds the occupation of planting to be the very soul of southern identity. Planters were caught up in a "tobacco metality" in colonial times that flowed from the very act of cultivation. See T. H. Breen, "The Culture of Agriculture: From Tobacco to Wheat in Tidewater Virginia, 1760-1790." Paper read before the meeting of the American Historical Association, Washington, D.C., December 31, 1981; Janet Sharp Hermann, *The Pursuit of a Dream* (New York: Oxford University Press, 1981), pp. 6-9; Aubrey C. Land, "Economic Behavior in a Planting Society: The Eighteenth Century Chesapeake," *Journal of Southern History* 33 (Autumn 1967), pp. 467-85.

17. F. L. King to Linn, February 17, 1865, Thomas Butler King Papers, Southern Historical Collection, University of North Carolina Library, Chapel Hill, North Carolina.

18. The most sophisticated elaboration of this approach is that of Eugene Genovese as found in his *The World the Slaveholders Made* (New York: Vintage Books, 1971). Genovese suggests that slavery affected the planter's desire to seek out and maximize profits by promoting a master-class world-view that involved aristocratic rather than bourgeois tendencies. This, by the way, is not the same point made by Woodman in commenting on Rothstein's "Hypothesis" in the above. Woodman feels that historians should not be surprised that southern planter elites seldom directed their economic activities into the home market, since to do so would have threatened their economic and social well-being by disrupting the status quo. Although an important insight for dealing with the planter as the agent of outside economic interests, Woodman's point does not detract from the possibility that the nature of the status quo rendered the South's antebellum planter elites incapable of regarding economic activity, inside or outside of the home economy, in other than aristocratic terms. Outside investments were clearly resorted to in a large way, but it is not clear that they were pursued. See also the perceptive statement describing the South's broad mixture of plantation master types, ranging from Nabobs to profit grubbers of the crudest sort, in James L. Roark's *Masters Without Slaves: Southern Planters in the Civil War and Reconstruction* (New York: W. W. Norton & Co.; 1977), pp. 68-108.

19. U.S. Census (1860), Manuscript Population, Slave, and Agricultural Schedules, Adams County, Mississippi, and Concordia Parish, Louisiana.

20. Ibid.

21. See William Kauffman Scarborough's *The Overseer: Plantation Management in the Old South* (Baton Rouge, La.: Louisiana State University Press, 1966) for a succinct and important analysis of the southern overseer's managerial duties and responsibilities.

22. U.S. Census (1860), Manuscript Population, Slave, and Agricultural Schedules, Adams County, Mississippi, and Concordia Parish, Louisiana.

23. Scarborough, *The Overseer*, pp. 102-137.

24. J. A. Gillespie Papers, Louisiana State University, Baton Rouge, Louisiana; U.S. Census (1860), Manuscript Population, Slave, and Agricultural Schedules, Adams County, Mississippi, and Concordia Parish, Louisiana.

25. U.S. Census (1860), Manuscript Schedules, Adams County, Mississippi, and Concordia Parish, Louisiana.

26. See Merle C. Nutt, *The Nutt Family Through the Years: 1635-1973* (Phoenix, Ariz.: Merle C. Nutt, 1973), pp. 93-137.

27. C. Peter Ripley, *Slaves and Freedmen in Civil War Louisiana* (Baton Rouge, La.: Louisiana State University Press, 1976), pp. 146-59.

28. Report of Colonel Samuel Thomas to Brigadier General Lorenzo Thomas, June 15, 1864, Records of the Adjutant General—Colored Troops Division, Record Group 363, National Archives.

29. Ibid., Manuscript Tax Rolls (1818-1861), Adams County, Mississippi.

30. James, *Antebellum Natchez*, pp. 183-84.

31. Ibid., pp. 156-57. See also Paul W. Gates, *The Farmer's Age: Agriculture, 1815-1860* (New York: Holt, Rinehart & Winston, 1960), p. 148; U.S. Census (1860), Manuscript Population, Slave and Agricultural Schedules, Adams County, Mississippi, and Concordia Parish, Louisiana.

32. Manuscript Tax Rolls (1818-1861), Adams County, Mississippi; U.S. Census (1860), Manuscript Population, Slave and Agricultural Schedules, Adams County, Mississippi, and Concordia Parish, Louisiana.

33. *Natchez Courier*, October 15, 1850, p. 2.

34. Appendix A.

35. Gavin Wright, "Economic Democracy and the Concentration of Agricultural Wealth in the Cotton South, 1850-1860," in William N. Parker, ed., *The Structure of the Cotton Economy of the Ante-Bellum South* (Washington, D.C.: Agricultural History Society, 1970), pp. 63-94.

• 3 •

"WITH THE HELP OF THE FREEDMEN'S BUREAU . . ."

The outbreak of the Civil War and the subsequent presence of Union troops as an army of occupation in the Mississippi Valley meant a new relationship between whites and blacks in the area. As a vanquished foe, former slave masters found themselves bankrupt in some instances, the owners of devastated property in others, and slaveless in all cases. Their former chattel became refugees first and then, with the Emancipation Proclamation, free men with labor to sell or to employ on their own account. Although potentially a revolutionary situation, the outcome of it all was a system of land tenure for southern blacks characterized by abject poverty, dependence upon planters and merchants, and a caste rigidity that bordered on peonage. By 1889, sharecropping (which was both a system of wage labor and a form of tenancy) had emerged as the dominant mode of agriculture and rural society in the South. The typical black farmer lived on rented lands in cabins owned by landlords who were often merchants. He grew only cotton, of which he either received a share in payment for his wages or else paid out a share as rent. In either case, he was usually in debt to the store, which was often owned by his landlord. His children worked on the cotton patch and possibly supplemented the family income by hiring out for cash wages at cotton-picking time. Regardless of their status (either as share wage-earners or as share renters), few of them had much say in the crucial decisions of farming or managed to escape the poverty of their lives. Although the system was far better than slavery in respect to their personal freedoms, their actual geographic mobility, and the security of their families, there is no question that the system was an intolerable burden upon southern blacks.

Sharecropping did not emerge overnight. In the Natchez District, years were to pass before the system was firmly in place, and even then it was seldom recognized as an institution of long-lasting duration. Between the death of slavery and the birth of sharecropping, the U.S. Army and the Freedmen's Bureau largely determined the freedman's role in the district. As the actual liberator of the slaves, the U.S. Army found itself responsible for protecting, keeping, and caring for the area's emancipated refugees. The duty touched on every aspect of the freedman's condition, but especially forced the army to confront the effect of their guardianship on the future of blacks as a free people. A decision had to be made as to whether the displaced freedmen were to be schooled and made ready for their freedom or simply cared for as refugees.

After the war, the army's responsibility shifted to the newly created Freedmen's Bureau, an agency especially designed to safeguard and promote the freedman's newly won freedom and equality. Never fully able to take the steps necessary for securing the former slave's social and economic independence, the bureau instead worked to at least protect the rights of blacks as free wage laborers. In the process, in the North's identification of freedom with wage labor, the stage was set for the emergence of sharecropping.

THE ROLE OF THE UNITED STATES ARMY

The Natchez District suffered less during the war than other parts of the South. The early and relatively easy taking of Natchez and its surrounding countryside by Union forces in the autumn of 1862 kept wartime destruction in the area to a minimum. When General T.E.C. Benson steamed upriver from New Orleans to cut off the flow of Texas cattle into Mississippi at Natchez, he found only a small resistance force in the town, and a few well-placed shots sent the enemy fleeing into the countryside.[1] The fact that Natchez fell with little difficulty is partly explained by the Confederate decision to concentrate its strength at Vicksburg and partly by the pro-Union sentiments held by numerous planters in the district. Adams County, a Whig stronghold before the war, had delivered one of the largest pro-Union votes in the state's secession referendum in 1860.[2] Although few planters in the district enjoyed having their neighborhood occupied by Union Soldiers, it was not complete hypocrisy on their part to welcome Union forces with waving handkerchiefs and unfurled American flags. For some Natchez planters, this was not a war of their making.[3]

But the war did bring economic and social upheaval to the area. Few planters managed to carry on their plantation enterprises during the war years. On the one hand, Confederate marauders, using the Louisiana swamps as refuge, terrorized, abducted, and even killed the work force of loyalists and northern speculator-planters, while, on the other hand, the federal government's policy of confiscating baled and standing cotton until certain of its owner's loyalty greatly complicated the situation. In addition, General Ulysses S. Grant's daring sweep through the Louisiana swamps to lay siege to Vicksburg from below, destroying levees and dikes in the process, threatened with ruin the rich cotton lands in Concordia parish and the lower parts of Adams County. Although left vulnerable to the Mississippi River's devastating spring overflows, the area escaped flood damages during the war itself.[4] But the threat was real enough. Indeed, massive flooding ravaged the area in 1866, 1867, and 1868, and a generation passed before the district's antebellum system of dikes was fully rebuilt.[5] As a result, the possible economic advantage of being occupied so early in the war never materialized. Even the healthy demand for cotton failed to draw resident planters into risking their remaining capital. Rather, district planters generally preferred to lease their lands to those outside speculators who were willing to undertake the risks of planting lands unprotected by levees and dams.[6]

The war witnessed the flight of thousands of slaves to Union lines in what amounted to a deliberate, if often cautious, break for freedom. Few of the refugees knew what to expect. William Faulkner's fictional depiction of an unrelenting wave of people pressing ever onward to the promised land, impervious to all hardship with freedom so near, is apt but possibly misleading in the sense that few of them wandered aimlessly.[7] In the Natchez District, for instance, the fleeing slaves headed towards the strongest Union garrison in the area, the town of Natchez. This town had been accessible to them each year on Christmas Day when the most faithful were allowed to visit there to enjoy a kind of freedom that gave Natchez a special meaning in their lives. Natchez was the most logical destination for the fleeing slaves.

The flight itself was significant for many reasons, but it deserves our attention because it forced the U.S. Army to regard blacks as refugees. Within days of the town's fall, General E. G. Ransom, the ranking officer in the Natchez area, reported the situation out of control. Slaves were flocking to.Natchez "by the thousands (about one able-bodied man to six women and children)," and Ransom hoped that orders would direct him to send them on to Vicksburg—a relatively easy task in his mind since "they are all anxious to go, though they do not know where or what for."[8]

Although the rush of slaves to Natchez confused the army personnel in the area, the high command in the Mississippi Valley had already reached several policy decisions on the refugees. These decisions came to serve as guidelines for the army's treatment of blacks during and after the war. Grant's campaign for Fort Donelson in Tennessee had presented the general with problems similar to those Ransom was to experience several months later in Natchez. After Grant's victory at the mouth of the Cumberland River, thousands of refugees trailed after his victorious army, seriously limiting the general's flexibility in the field.

Grant, of course, viewed the refugee problem mainly from the standpoint of military necessity. While understanding that a mass of fleeing slaves might well demoralize the enemy, he was especially disturbed by the thought of deploying valuable resources in looking after the straggling, starving refugees. Nor did the general ignore the effect his role as the black's official guardian would have on troop morale and the precarious loyalty of so-called neutral planters. But even Grant's singular military outlook was tempered somewhat.when faced with the obvious suffering of so many displaced people. As a result, he worked to mesh into one policy the problems of considering the blacks as (1) contrabands of war to be used against the enemy, (2) counterproductive to the army's efficiency in the field, and (3) a suffering humanity.[9]

Hoping to press on to Vicksburg after his victory at Fort Donelson, General Grant instructed Colonel John Eaton, a chaplain in the Fifth Regiment, to organize the contrabands in work companies to pick, gin, and bale any cotton ungathered on the abandoned plantations in the vicinity. Grant thus hoped to discipline and protect the refugees in a swift and orderly fashion while allowing them to pay as much of their own way as possible. The cotton picked belonged to the U.S. government, but the proceeds would at least pay for the rations consumed. Once the abandoned cotton was picked, the plan called for employing the refugees as woodchoppers and in general fatigue labor for the army. Stationed along the river and in backup details, refugee slaves might thus perform much of the maintenance work around the Union camps in order to free white soldiers for fighting.[10]

Grant's primary concern with the military problems and possibilities posed by the refugees remained the guiding principle shaping the army's policy towards the freedmen during the war years. Unless this fact is understood, the army's role in the origins of sharecropping is needlessly confusing. But it is also important to note that the refugee situation involved complexities that substantially modified and competed with the military needs of the moment. These often conflicting realities included short-range concerns for individual profits by army officers, loyal planters, and northern speculators in the area; political and personality clashes among the men in authority; the ideological concerns for the long-range effect of policy upon the freedmen; and the moral concern for justice by abolitionists and sympathetic northerners. Military goals usually determined the outcome of policy, but other factors were always involved.

The fact that Grant appointed a chaplain as superintendent of the contraband with absolute authority over their welfare is indicative of the general's feeling that more was at stake than simply organizing the refugees into work camps for military needs. Along with the short-term but all-important military goals of the moment, Grant believed that the army's task involved the delicate and difficult chore of teaching the slave refugees to act as responsible, hard-working individuals. Feeling certain that the refugees would soon be emancipated and required to prove their worth to the country by laboring diligently and, if they worked well, by fighting for their freedom as soldiers, Grant hoped the army might perform a lasting service to the country and the refugees by instilling in them a sense of responsibility defined in terms of the classical American work ethic: faithful labor for just wages. The freedmen's future, he felt, depended on their initial performance as wage laborers.[11]

Accordingly, Grant initiated a policy, through Eaton, which was designed to put the refugees to work in the cotton fields under government auspices. The labor contract was employed as the instrument most likely to achieve his goal. Every able-bodied adult slave was ordered to labor for fixed wages at picking cotton that was still standing on the abandoned plantations. In the process, Union officers would instruct the workers in the meaning of contract labor, answer questions, and oversee the signing of contracts among those slaves remaining on their masters' places. If all worked according to plan, the army would become the chief employment agency in the Union-held areas. Neutral and even disloyal planters would either have to employ blacks on terms set by the military or give up farming for the time being.[12]

Grant's policy was meant to function in a curiously one-sided fashion. Wages were set at subsistence levels, with the earnings either channeled into a general fund supporting the entire refugee camp (in the case of the government operations), or else paid out in food and clothing rations as in slave times (in the case of the privately run plantations). The contracts also allowed for working the refugees and resident slaves in gangs as had been the case in slavery. They had no real part in establishing either the conditions of labor or the terms of contract, and many of the rules regarding discipline and control which had operated before the war continued. But Grant thought the slaves would now be selling their labor for wages and would therefore benefit by learning the meaning of a contract.

Even though Eaton worked to implement Grant's instructions as he understood them, the program never functioned as the army had hoped. The very success of the federal movement into the Mississippi Valley in 1862 and 1863 undermined the army's ability to handle the situation. Thousands of blacks poured into the Union camps once Vicksburg fell, and the Emancipation Proclamation forced the army to liberate slaves from deep within enemy lines.[13] Faced with thousands of propertyless blacks, Eaton's skeleton program floundered in a sea of human suffering. Refugee camps initially set up to put the blacks to work gathering unpicked cotton or else in fatigue labor for the army instead became disorganized barracks where the fleeing slaves found only disease and barely subsistence fare. What had begun as an optimistic experiment to aid the war effort while preparing the slaves for freedom was instead reduced to the mere essentials of care and protection.[14]

But Washington soon followed up Lincoln's Emancipation Proclamation by sending the adjutant general of the U.S. Army into the Mississippi Valley with full authority to supervise the refugees. General Lorenzo Thomas agreed with Grant that the refugees should bear the burden of their upkeep by working. Nonetheless, he introduced several new features with long-range effects. In the first place, Thomas carried orders to begin drafting black males into a home guard of colored troops led by experienced white officers. Trained and equipped as fighting units, these black companies of men were to take an active role in policing and guarding the countryside, thus freeing regular army units for duty elsewhere. Second, Thomas planned to lease abandoned plantations to loyal whites who were willing to work black refugees for fixed wages. These two elements of policy, in Thomas's opinion, complemented one another. With most abandoned plantations near or adjacent to the Mississippi River, a loyal population so strategically placed would greatly lessen the army's defensive role in the area. In addition, since the black army units would be literally defending their homes, they could be expected to perform most effectively because of the duty involved. Finally, although the federal government might continue to work some lands on its own account, the army's responsibility for supervising the refugees in their daily labor would be largely taken over by the lessees of the abandoned lands. Thomas hoped that the army could thereby limit its direct involvement to the black troops and to several so-called Home Farms, established to care for the sickly and aged and to provide centers where lessees could find laborers, while the vast majority of refugees would find work on the plantations as hired hands.[15]

Notably absent from Thomas's plan was much concern for its long-range effect upon the freedman's work ethic or future independence, although such considerations were not completely ignored. In Thomas's mind, the military security of the Mississippi Valley was of the highest priority. This was to be expected during war. But his neglect of the opportunity to promote long-term goals may have been based upon his personal commitment to preserving the class and racial status quo. Thomas had been stationed in the area as a young officer and knew many of the loyal and rebel planters on a first-name basis. Moreover, there is evidence that he had personal investments in the neighborhood which were best secured by limiting the economic independence of the refugees as free men.[16] All this is mere speculation, although it is clear that Thomas's program strikingly excluded Grant's and Eaton's more paternalistic concerns for the freedman's ultimate place as a free, independent citizen, a paternalism rooted in their feelings that blacks could in time become something more than serfs.

Thomas's plan did call for a continued role for the Eaton superintendency. Briefly, Thomas selected officers known as provost-marshals from the men in Eaton's department to function as superintendents of the freedmen in the Mississippi Valley. Charged with the duty of pro-

Plate 4. Battle of Milliken's Bend.

tecting the freedmen and representing the federal government, these officers patrolled the leased plantations under the authority of military law. Previously, officers known as provost-marshals had functioned as the police in Union-occupied areas, sharing with the agents of the U.S. Treasury in the responsibility for collecting rents and taxes. The Thomas policy combined the duties of the policeman, landlord, freedman's agent, and government representatives into one office, with Eaton remaining, under Thomas, as the superintendent of freedmen in the valley.[17]

Eaton began Thomas's program of leasing plantations in April 1863, early enough in the season to clear the fields and make ready for planting. He set rents in the form of a tax of $2 per bale of cotton and 5 cents per bushel of corn or potatoes produced.[18] Although the military never guaranteed protection, most planters in the area assumed that troops were available, patrols organized, and definite plans for a regular home guard of black troops in the making. In return for this protection, the loyalists pledged to hire blacks, furnish them with rations of food and clothing at cost to be deducted from wages due, and pay the relatively low scale of $7 a month for adult males, $5 for women, and $3.50 for all hands ages twelve to fifteen. In addition, the hired workers would pay a tax based on wages earned for the support of sick and dependent refugees.[19]

But Thomas's plan met with difficulties from the start. In the first place, only a small portion of the total abandoned lands, mainly those lying close to Vicksburg, was leased in 1863. The problem stemmed largely from the army's inability to protect the plantations from Confederate raiders operating out of the Louisiana swamps.[20] As a result, hundreds of blacks in a vast area from Baton Rouge to above Natchez were herded into refugee camps and cared for by the army on a rations basis or else allowed to do as they wished. It was in these areas that many blacks joined together and began planting corn and potatoes or a little cotton with the full approval and encouragement of the military. In the second place, the unscrupulous activity of northern speculators so demoralized the refugees that many refused to return to the fields. The extent to which speculators paid blacks in the cheapest furnishings at inflated costs or skipped out in midseason without settling wage and rent bills is well documented. Less known is the frequency with which plantations were leased and refugees hired by men who had no intention of ever planting cotton. The records of the War Department are filled

Plate 5. "Driving them off the Plantation without Wages and Shooting them."

with instances of lessees renting plantations as fronts for their illegal business of gathering up cotton abandoned in the countryside or else procured from agents who hoped to evade paying federal taxes. Moreover, agents of the federal government conspired with unscrupulous leaseholders and freedmen in intricate plots for practicing fraud upon both the refugees and white planters in the area. More than a few officers later faced court-martial charged with taking bribes, supplying plantations from military depots in return for payoffs, and forming partnerships in conspiracy against the interests of government and freedmen alike.[21]

To be sure, fewer instances of fraud and unsettled accounts occurred when soldiers were readily available to ward off enemy raids and to keep the lessees in line. But the degree to which the army was negligent in this matter came to light in William Yeats's report to the U.S. Treasury in October 1863. Yeats, after touring the Mississippi Valley as the agent of an abolitionist organization, charged the army with gross inefficiency and mismanagement in handling freedmen's affairs. In his view, the heart of the problem stemmed from leasing lands to speculators and northern capitalists with too few safeguards built into the system. Using Yeats's report as the major block from which to build, the U.S. Treasury successfully moved to gain control of the whole system of plantation management. The department had long defended this position on the grounds that leased plantations, as a source of revenue to the federal government, fell under its financial and administrative policies. Within weeks, the system was turned over to the treasury, although the army retained authority over refugee camps and the affairs of freedmen in general.[22]

Accordingly, Yeats, along with a special agent of the Treasury Department, William P. Mellon, was assigned to reorganize the Thomas program in the Mississippi Valley. Although very critical of the then existing provisions for employing the blacks, the Yeats-Mellon plan nevertheless endorsed the Grant-Eaton-Thomas concept of having the freedmen work for wages rather than on their own account.[23] The major revisions suggested by the Treasury Department had to do with wage rates and the stipulations governing the lessees. Briefly, no lessee was permitted to rent more than one plantation, and wages were increased to $20-25 for adult males, $18-20 for adult women, and $15 for those aged twelve to fourteen and over fifty. According to Yeats, the wage increase would enable freedmen to supply themselves and thereby relieve the Treasury of any cumbersome responsibility for supervising supplies. More importantly, it was hoped that the new situation would force speculators to abandon leasing in favor of smaller farmers as the exorbitant profits from the supply business and cheap labor disappeared.[24]

The Yeats-Mellon plan met with immediate opposition from Thomas and Eaton. Faced with the continued responsibility of caring for indigent blacks in the Home Farms, Thomas charged that the new policies forced more freedmen into the camps then ever before, since few lessees were willing to hire any but the best hands at such high wages. The sick and elderly were cast out of families, capital fled the country causing massive unemployment even in the safe areas, and the removal of the plantation management from army control meant less protection for outlying plantations. Not to be overlooked, according to Thomas, was the fact that by allowing freedmen to supply themselves the treasury had encouraged unprecedented looting and stealing throughout the lower Mississippi Valley.[25]

The intensity of the charges exchanged between Yeats and Thomas finally brought President Lincoln's personal intervention in favor of a compromise. While returning control of all affairs involving the freedmen to the army, the president ordered Thomas to work with Yeats in reforming the inequities of the earlier system. Losing little time in carrying out the president's order, in March 1864 Thomas issued the· most complete statement to that date defining the role of blacks during the war. Order Number 9 rested, however, upon several features present in the army's dealing with the refugees since Grant's initial policy. For one thing, Thomas explicitly accepted the idea that southern blacks must work as cultivators of the soil and implicitly embraced the theory that, while some freedmen should be encouraged to work as tenants on their account, most would function best as wage laborers. Accordingly, the army's main responsibility was to prepare the freedman "for the time when he can render so much labor for so much money, which is the great end to be attained."[26]

But much of the order attempted reforms along the lines suggested by Yeats and Mellon. Clearly specified for the first time were the various levels of responsibility: those of the army, employers, and freedmen. Promising to utilize the military to the fullest extent possible in protecting the leased plantations, Thomas vowed that the army would never

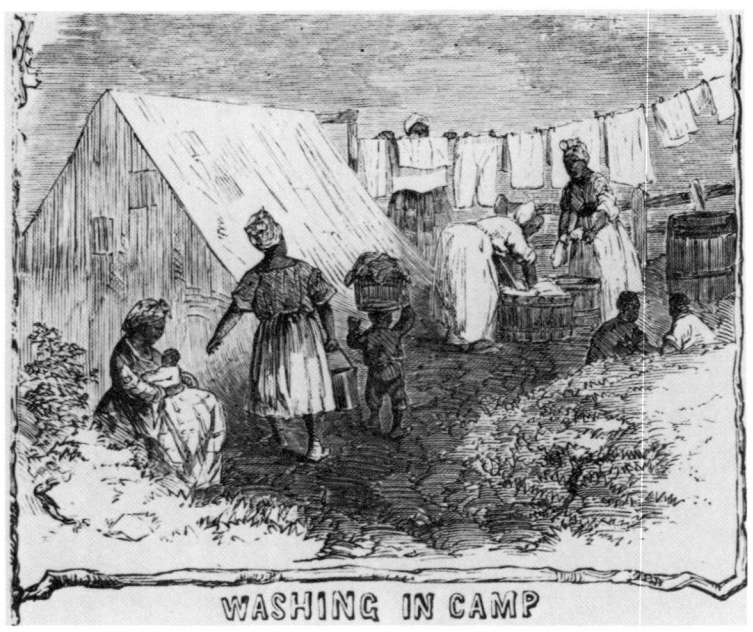

Plate 6. Washing in Camp.

again seize boats or wagons or any other supplies used in plantation work. Nor would any future army personnel be allowed to enlist plantation laborers as soldiers. Black soldiers, a source of constant irritation to overseers and planters, were absolutely forbidden to visit plantations without written consent from the planter and never, in any case, bearing arms. In terms of administration, all leased plantations were again placed under military authority in general, with the various provost-marshals responsible for the enforcement of justice and equity between the freedmen and whites.

Regarding the responsibilities of employers, the order decreed that although lessees might once again rent more than one plantation, under no circumstances could they purchase clothing or other products from any laborer or "commute supplies, except clothing, at the rate of more than $3 per number." Work on Sunday entitled the hands to overtime pay, and it was suggested that employers provide certain freedmen with the opportunity to cultivate lands on their own account to encourage independence and self-reliance. In the latter case, Thomas was only endorsing the accepted theory that a few blacks might actually work themselves up the agricultural ladder to a kind of tenancy status rather than establishing a recommendation that might apply to large numbers of freedmen. Finally, Thomas insisted that planters should employ only those overseers who accepted the "new situation" existing between a planter and his laborers. Freedmen were no longer slaves and must now be treated as hired hands rather than chattel.[27]

As with previous procedure, Order Number 9 saw the Home Farms of the government as only temporary expedients until the day when freedmen were fully employed as hired hands on privately owned or leased plantations. As such, Thomas defined the freedman's responsibilities in terms of good and faithful labor, thus reflecting the theory that slavery had left blacks unprepared to function as free laborers responding to the incentives of wages by working long and hard and well. It was therefore absolutely necessary to give the employers enough control over labor to protect investments but also to keep the freedmen working long enough to train them as disciplined field hands. Only then could the dual goals of military security and the preparation of blacks as wage-conscious employees be achieved.[28]

The problem was to keep the blacks working. In the first place, jobs had to be available, which meant making farming attractive enough to entice northern capitalists to begin leasing farms and planting. Thus, wages were reduced from Mellon's high down to $10 per month for adult males, $7 for adult females, and half the above amounts for children ages twelve to fourteen, with rations, clothing, and quarters supplied by the employer. Order Number 9 established several means by which the employer might be guaranteed a good day's labor from his hands. Specifically, wages were to be paid in set portions on a monthly basis, with one-half held by the employer until the season's end to insure that the laborers worked out the entire contract year. This precaution was in direct response to the planters' often voiced complaints that hands paid on a weekly or monthly basis frequently abandoned the crop at picking time. In a similar vein, Thomas forbade blacks to leave the plantations unless authorized, made contracts that were binding under military law upon employer and employee alike, and promised to punish vagrancy by putting the unemployed to work on public works (roads and levees).[29]

"Respectful, honest, faithful labor" meant more than a hand's presence on the place through picking time: it especially meant obedient labor. In former times, the lash had served this purpose, although the shrewdest planters always employed other incentives as well. But with "flogging and other cruel or unusual punishment" interdicted, the only recourse left was in the employer's control over wages and livelihood. Accordingly, disobedience, insolence, and time lost because of illness would be the hand's loss and not the employer's. In minor cases, the problem would be solved by wage forfeiture, but in cases of "attempted imposition, by feigning sickness or stubborn refusal of duty, they [the Negroes] will be turned over to the Provost-Marshal of the police district for labor upon the public works without pay." The freedmen were thus expected to work as well as ever, "between daylight and dark," because labor was a "public duty and vagrancy a crime," but chiefly because they were still that "people identified with the cultivation of the soil, however changed in condition by the revolution through which we are passing." In a word, in Thomas's opinion, emancipation had not freed the black from his class responsibilities.[30]

Although Thomas did recognize the possibility of contracting with hands on a share basis in his Order Number 9, the procedure was never considered a viable alternative to wage labor. Indeed, Thomas never deviated from the recommendation of the Western Freedman's Inquiry Commission that fixed cash wages were both the cheapest and most just remuneration for the freedman's labor. Rather, he thought of shares as a means of providing for "the current wants" of the wage hands by appropriating lands for share cultivation. According to this proposal, the hired freedman might receive a plot of land for growing corn or even cotton on his days off, with seeds and tools and animals possibly rented from the employers for a share of the crops made. In this sense, the share system was viewed as complementing the fixed wage system by allowing the freedmen partly to supply themselves. There is no evidence that Thomas thought of the system as either a form of wage settlement or tenancy for the freedmen refugees.[31]

Thomas's Order Number 9 failed to bring much stability to the scene. Disputes between the army and the treasury continued throughout 1864, with both groups sharing responsibilities and haggling to the ultimate confusion of all. But most damaging was the army's continued inability to protect adequately the plantations, especially those leased on the west side of the Mississippi River. When such circumstances combined with the damage done by flood waters careening over makeshift dikes and levees, mere proclamations of policy failed to stop the blacks from abandoning the plantations for the army camps and towns.[32]

In brief, although the army's official policy during the war years varied and was seldom fully implemented, it viewed blacks primarily as fixed wage agricultural laborers. The problem of how to keep the lands occupied and the freedmen working for reasons of military objectives dominated policy. This concern led naturally to a preference for staffing the plantations with gang, low-wage laborers as an incentive to attracting northern lessees to the neighborhood. Such continuity between the working conditions of slavery and the refugee experiment was an easily policed program, low in cost, and of limited threat to the status quo of loyal planters in the area. At no point did either Grant, Eaton, or Thomas seriously consider establishing a general system of tenancy as a viable alternative to wage labor on a fixed basis, although Eaton had experimented with a limited notion of "self-directed labor" for "exceptional" blacks at Davis Bend above Vicksburg.[33] Tenancy and its concomitant small-family-farm status for refugees was apparently too much of a social revolution to be taken up, besides being economically and militarily unfeasible. Few northern capitalists would have furnished enough credit to make the concept a reality in the war-torn Mississippi Valley, and the problem of militarily protecting thousands of tenants was an impossibility. In its place, as the best solution to the refugee problem, the army favored fixed wages and the status of field hands laboring in gangs for subsistence.

THE FREEDMAN'S BUREAU

When Congress created the Freedmen's Bureau in March 1865 and placed it under the jurisdiction of the U.S. Army, it seemed that the Grant-Eaton tradition would continue. The selection of O. O. Howard, a successful Union general, as commissioner of freedmen, and Howard's subsequent appointment of several officers from Eaton's headquarters as assistant commissioners were hardly precedent-breaking. But while the staff continued largely intact, overall policy underwent significant changes. In the Mississippi Valley, as elsewhere, the Freedmen's Bureau found itself committed to undertaking a radical program for implementing the congressional directive to distribute abandoned lands to the freedmen to the amount of 40 acres per family. Although not enough confiscated lands were available to give every black family a farm, it was hoped that enough could be distributed to establish a precedent for the ultimate parceling out of lands to freedmen. Howard thus selected his assistant commissioners on the basis of their willingness to implement this revolutionary order. Unlike the Thomas policy in the Mississippi Valley, Congress had decided to endorse the more radical experiments associated with the South Carolina Sea Island model. Under this model, blacks were given land to work on their own, not as wage hands or tenants but as free men owning homesteads.[34]

President Andrew Johnson's veto in September of a bill extending the bureau's life, although overridden, put an end to the agency's most radical stage. What followed, even during the heights of Radical Reconstruction, was a scaling down of the bureau's ability to assist in a true social and economic revolution from the ground up for the freedmen. The most radical of the assistant commissioners found themselves generally forced out and replaced by men who were more favorable to the interests of the southern planting class. Once the entire operation merged with the army under the first Military Reconstruction Act in March 1867, the question of confiscation and subsidies for the freedmen was a closed issue. The bureau's official position in the Mississippi Valley was thereafter one of protecting the freedman's right to secure for himself the just fruits of his labor, that is, wage labor.[35]

Although denied the role of fulfilling a revolutionary program, the leeway exercised by the bureau's various assistant commissioners largely determined the freedman's progress from slave to farm worker. The Natchez District fell under the authority of Colonel Samuel Thomas, the assistant commissioner of freedmen in Mississippi. With his headquarters at Vicksburg, Colonel Thomas served in this capacity until 1867 when he was replaced as assistant commissioner of the Freedmen's Bureau by a General Thomas Wood. Wood was himself replaced in 1868 by General Alvin C. Gillim. Although their authority included more than the Natchez District, the bureau's role in the neighborhood is largely the story of the first of these officers.

Colonel Samuel Thomas had served as John Eaton's right hand since 1863. His devotion to the cause of freedom and equality for the refugees stemmed partly from his abolitionist sentiments but more importantly from his experience as the commanding officer of a black regiment during the war. As a former provost-marshal, he also had first-hand experience in dealing with the tenuous relationship of black people to the white community.[36] It is not surprising, therefore, that Thomas accepted with optimism his appointment as assistant commissioner of the bureau in Mississippi. But President Johnson's opposition to the distribution of lands, on the one hand, and the colonel's inability, on the other hand, to gain clear authority over the management of abandoned lands from the Treasury Department hemmed in the young officer from the beginning. His hope of giving freedmen the opportunity to rent lands with the option to buy in three years was crushed by Johnson's amnesty policy.[37]

Unable to secure much land for the refugees to farm on their own, as an alternative and at the direction of Howard, Thomas pursued the policy of enforcing strict adherence to the contract system. Thomas was acquainted with the freedman's difficulty in obtaining wages from speculators during the war or in making enough to get a grubstake under the company store supply policies of the low-wage leasing system, and he was also clearly aware of the southern whites' broad desire to reenslave the blacks if possible. He therefore instructed his agents to supervise the writing of contracts with an eye to justice and the needs of freedom. Fair wages for a fair day's work would eventually secure for the freedmen the capital necessary to lease and then buy their own lands independently and freely. Accordingly, since it appeared doubtful that the Freedmen's Bureau would be able to distribute land after Johnson's amnesty proclamation, the bureau worked to persuade, indeed to force, freedmen to labor diligently on the plantations for what Thomas believed was their own good.[38] Such reasoning also lay behind Thomas's enforcement of the state's vagrancy laws. He considered his support of the apprenticing out of black orphans to planters to be the lesser of two evils because the alternative was starvation. Similar reasoning lay behind his policy of transferring black laborers from interior counties to the Mississippi River country to work for wages on the leased plantations.[39]

Convinced that the law of supply and demand rather than a high minimum wage would best enable the freedmen to secure the highest wage possible because of the relative scarcity of labor in the valley, Thomas, breaking with procedures established by Eaton and General Lorenzo Thomas, refused to set wages.[40] Nor did he believe in establishing rules for the contracts negotiated. Rather, in his opinion, contracts should result from bargainings between employer and employee, and, of course, the employee was advised by the bureau's agents. Only then might (1) freedmen learn to negotiate to their best advantage and (2) capital retain its traditional innovative capacities.[41] Thomas rejected enforcing hard lines designed to guarantee the freedmen high wages and sure payments. He instead limited the bureau to counseling and arbitrating disputes in a fashion similar to the old provost-marshal's police power on the plantation. In this context, planters generally found it to their advantage to have agents of the bureau nearby to settle disputes. Several officers under Thomas's command actually received retainers paid by planters in the neighborhood.[42]

There is little reason to doubt that Thomas sincerely believed these steps would insure freedmen at least the chance of earning the grubstake capital necessary for working lands on their own accounts. In fact, once it became clear that planters had used the apprentice concept to exploit and in some instances enslave black children (many of whom had parents capable of supporting them) and that speculators had brought freedmen laborers into the valley under promises never fulfilled or in conditions that broke up black families, the colonel repudiated both measures and vigorously acted to put an end to the situation as quickly as possible.[43] In one instance, Thomas clashed with the treasury in an unsuccessful effort to increase the rents on leased plantations so as to reduce the number of speculators in the valley.[44] At another time, he candidly confided his pleasure to General Howard that the freedmen "won't work as in days gone by." "Ten hours a day will educate his children and support himself if he is not cheated. I am glad that they [the freedmen] refuse to work two more hours merely to add a few dollars to the income of his [their] white lord."[45] When the bureau learned that whites in the area could not be coerced into leasing or selling lands to blacks, Thomas quietly supported several efforts to buy lands for them through white agents—in some instances, former army officers in the black regiments.[46]

Here then was the dilemma of a well-intentioned man: convinced that the freedmen should be working on their own lands or renting, else they might lose their chance at true independence and freedom, yet unable to do much about it directly, Thomas found himself promoting the interests of the planter-landlords in establishing a suitable work force under wage labor conditions. His decision to allow for varied working arrangements and freedom in wage terms forced the bureau officially to take a back seat in the actual terms of contract. Although no official policy dictated the form of the labor arrangements for blacks, Thomas and his agents indirectly promoted the so-called share notion of wage settlement and tenancy. It did so by encouraging, on the one hand, the leasing of lands to freedmen—a situation that could be accomplished only by relying on the share arrangement (without property or other creditworthy assets, freedmen could only offer the fruits of their labor as rent payments) —and by demanding, on the other hand, that laborers be given "a share of the place" whenever employers failed to pay wages due.[47] Unfortunately, the bureau's influence here is clouded in obscurity. Although the bureau's loss of authority over abandoned lands surely reduced its influence to little more than rhetoric and impractical arrangements in leasing or share tenancy, its role in regard to a share of the crops made as a form of wage settlement is les certain. Was the bureau's policy of insisting on the laborer's right to hold a lien on the crop for wages to be settled by "shares" the first step in establishing sharecropping? The answer is uncertain largely because the information on the number of liens enforced is unknown. Major G. D. Reynolds, for instance, the bureau's agent at Natchez, reported that freedmen refused to sign contracts that were not endorsed by the bureau's agents. Reynolds listed 2,298 freedmen in Concordia Parish who had registered contracts with his office in 1866.[48] Although certainly not the total number of freedmen in the parish, the number registered suggests that the bureau had significant influence in the freedman's everyday working affairs. But registration was not the same as the strict enforcement of wage settlements. Yet, it is likely that the bureau's continued insistence on the laborer's right to a lien on the crop for wages associated the share basis of settlement with justice, while linking the old fixed wage system with something less desirable. In both instances, however, the crop was viewed as belonging to the worker as a right of his labor. In any case, the bureau's role in the move from slavery to sharecropping under Thomas was less than direct.[49]

The primary assumption behind Thomas's policy was that freedmen must work on the cotton plantations in one capacity or another to secure capital and thus become independent in the long run but not charity cases in the short run. Such a policy clashed on several levels with the freedman's belief that freedom at least meant the right to reject working conditions associated with slavery. In the days immediately after the Emancipation Proclamation, freedmen tried the whole gamut of work possibilities open to free men. Large numbers of freedmen, looking for work and fleeing the intolerable swamps, spilled into the towns and villages at their first opportunity. Few towns in the valley were spared the sight of vagrant blacks loitering around the market or setting up their shanty towns at the edge of main streets.[50] But it would be a mistake to think that all of these people survived on charity or by stealing. Major Reynolds issued 869 city passes to freedmen supporting themselves in March 1865, while sending to the Home Farm at Davis Bend only three hundred others out of a total black population of over five thousand.[51] What these people did for a living is unclear. Possibly they barely survived while waiting for some better opportunity, but they did not abandon their shanty homes willingly. It eventually required both a state law on vagrancy and the cooperation of the Freedmen's Bureau in its stringent enforcement before the greater parts of the shanty towns were broken up.[52]

Others attempted to subsist off the land by fishing, cutting wood for sale to river boats, herding a few hogs, or even squatting on unclaimed lands and raising enough corn to live. In April 1866, the *Natchez Democrat* noted:

Few persons are aware of the extent of the Negro village situated on the river above the cotton press. There are probably not less than a thousand souls living there in small cabins, each surrounded by a little enclosure under cultivation. The cabins are equidistant and arranged with a degree of regularity which gives the place a very neat appearance. How these people live is a mystery to us, and yet the village seems to be in thriving condition. Unless the Mississippi River should inundate it, the inhabitants will doubtless produce this year a sufficiency of the products of the soil upon which to subsist. As meat is not being produced in that way, we think they will be a community of vegetarians.[53]

Still others entered the underworld of petty larceny and even semi-organized plunder in order to survive without going back to the plantations.[54] The disorganized state of affairs found convictions for cotton thefts and petty stock at an all-time high in Natchez in 1865—perhaps too high.[55] Indeed, Reynolds ordered Mayor William Dix of Natchez to halt the sale of property confiscated from the freedmen unless municipal officers had clear proof that the goods had been stolen.[56]

But the overwhelming majority of freedmen in the Natchez District willingly agreed to work as agricultural laborers primarily because they expected that the situation would soon change for the better. The evidence suggests that the hope for land homesteads of their own was a dream that made working for former masters or northern leaseholders acceptable to thousands.[57] Even when it became clear that no land was forthcoming, and in the face of the bureau's intensive efforts to convince them of this fact, the freedmen clung desperately to their dream. So obvious was their dismay as the truth sank in, that rumors of a massive uprising spread throughout the neighborhood in 1865. Although there is little evidence that organized revolution was brewing, thousands of blacks in the district refused to sign contracts in January so as to be unattached and available to take up the land and also in protest of the government's failure to keep its word.[58]

Colonel Samuel Thomas is remembered as one of the bureau's most radical assistant commissioners,[59] but all the same his programs varied only slightly from the Grant-Eaton-Thomas tradition. If he had been allowed to remain in his post with an adequate staff, this variation might have been important. Unfortunately, Thomas was recalled to serve in the Washington headquarters of the bureau in the spring of 1866, in a reshuffling of officers that coincided with the first Military Reconstruction Act, which in effect merged the office of assistant commissioner of the bureau with that of department commander. Thus, General T. J. Wood was made agent and chief army officer in Mississippi. Wood's appointment, along with that of his eventual successor, Alvin Gillim, indicates the extent to which Andrew Johnson retained control over the army during congressional Reconstruction.[60]

Wood, a native Kentuckian, and Gillim, from East Tennessee, were friends of Johnson and functioned primarily as administrators of army policy in dealing with the freedman issue.[61] Refraining from interfering with the existing civilian authority, Wood, and especially Gillim, generally encouraged a "cordial cooperation" between the military and the civil government.[62] Policy had been set regarding land distribution, and they believed there was no point in encouraging hope among the blacks for any immediate change. Wood sternly insisted on a policy of nonintervention into the affairs of the two races,[63] and, at his recommendation, President Johnson's secretary of state, Edward Stanton, disbanded black troops in Mississippi as a threat to peace.[64] Neither officer took little more than passing note of the charges that "regulators" were "plundering" freedmen in the interior or that planters were forcing freedmen to contract by the second Monday in February.[65]

Interestingly enough, however, during the Wood-Gillim tenure the Freedmen's Bureau in Mississippi for the first time began to encourage a new kind of working arrangement for blacks. In October 1866, General Wood explained to General Howard that "the present system of cultivating large plantations upon the wage basis might be improved upon by leasing in small parcels and making the freedmen responsible, in the crop, for supplies furnished them, leaving them the proceeds of the crop, after paying the year's advances and rentage as their pay."[66] Although Wood admitted that it was a novel or, as he put it, "undeveloped idea," he felt that the plan might be favorably received and thus help resolve a crisis situation that was partly of his own making.[67] In the first place, the general's policy of nonintervention into civilian affairs had enabled planter-landlords in the area to "demoralize" the freedmen so greatly as to pose the likelihood that many would refuse to work the plantations in 1867. For example, after having reluctantly signed contracts to work in gangs for people they mistrusted, numerous freedmen found themselves discharged in midseason for bad conduct in what was reported to be a shrewd move by planters to force the freedmen into forfeiting their wages due at year's end.[68] In addition, planters throughout the valley declared bankruptcy either because they wanted to avoid paying wages due or because their crops had failed in the wake of the disastrous spring floods and insect attacks of that year.[69] In the former case, the freedmen had little recourse, since the state's recently passed stay laws exempted land, tools, and appurtenances from confiscation for debts.[70] Out of work, because of fraud or natural misfortune, the suffering freedmen were thrown back onto the bureau.[71]

Within this context, sharecropping was recognized as a device that would at least enable the freedmen to secure the advances needed to keep them off government ration. Moreover, while Wood endorsed sharecropping as a form of tenancy, he spoke of the cropper's share as "pay" rather than income. Neither Wood nor his predecessor ever attempted to force the issue, and although the general later claimed that his suggestions "were being generally adopted by the planters," it is doubtful that the system originated on the strength of the superintendent's word alone. Rather, the general was undoubtedly proposing some-

Plate 7. Verdict, "Hang the D—— Yankee and Nigger."

thing that was acceptable for reasons that go beyond the bureau's role in the matter.[72]

CONCLUSION

The situation had therefore taken an ironic turn by 1866, for the bureau, under Assistant Commissioner Thomas, had functioned nearer its original radical design during presidential Reconstruction than after. The U.S. Army had supported fixed wages, annual contracts, gang labor, and the commitment to get the black refugees and resident population back to work on the slave plantations. The army's dual purposes were to provide, first, military security for the valley and, second, a training period in which blacks would grow accustomed to the fact that emancipation did not mean freedom from plantation labor. Although they were not to be slaves, neither were they to be free laborers who could themselves decide under what conditions they would work. Nor was there much encouragement for them to become self-directed tenants or landowning farmers.

But this is not to say that Lorenzo Thomas or his assistants worked to frustrate the development of such a farming status among the freedmen. The point is that military needs were primary, followed by the adjutant general's support for the loyal planter establishment. In his mind, except for the initial caring for the indigent refugees, any further paternalism should come from those who were most capable of extending it: loyal antebellum planters. It was not Thomas's goal to make serfs of the freedmen by making them dependent upon the army; this explains why he was hesitant even to allow for the paying of wages according to age classifications. Nor was it Thomas's goal to protect the freedmen from becoming the wage slaves of the plantation system. In confrontations with the Yeatses and Mellons of his world, however, he was eventually forced to give a clearer definition of the freedman's status as a worker in need of protection.[73]

With the Freedmen's Bureau, the army's policy of providing blacks with barely starvation wages shifted towards allowing market conditions to determine wages. Behind this shift was Colonel Samuel Thomas's conviction that the impoverished and undisciplined blacks could better free themselves from dependency by selling their labor to the highest bidder. According to his scenario, high wages would be the first step towards free farming status, the emergence of a mobile black labor force, and the possibility of true economic equality. As the alternative, Thomas saw the reemergence of slavery in the form of regulated wages, nonentrepreneurial habits, and manipulation. It was as if Thomas believed slavery had made the blacks unfit for a competitive economic system, on the one hand, and alone and without true protection, on the other hand. In this reality, Thomas believed blacks must be thrown into the water, learn to swim, and emerge from the experience a stronger and more independent people than before.

From the standpoint of the freedmen, their laboring conditions must have seemed absurdly beyond their control. At every turn, both their friends and their enemies were most concerned with having them work as disciplined wage laborers. This new status was to be their salvation.

NOTES

1. C. G. Dahlgren, Commandant, Confederate Post Washington, Mississippi, to Brigadier General Thomas Jordan, May 17, 1862, in *The War of the Rebellion: A Compilation of the Official Records of the Union and Confederate Armies* (hereinafter cited as *Official Records*) (Washington, D.C.: U.S. Government Printing Office, 1880-1901), Ser. 1, Vol. 15: 737-38; Brigadier General Thomas E. G. Ransom, U.S. Army, to Lieutenant Colonel W. T. Clark, Assistant to the Adjutant-General, July 16, 1863, *Official Records*, Ser. 1, Vol. 24: 680-81.

2. D. Clayton James, *Antebellum Natchez* (Baton Rouge, La.: Louisiana State University Press, 1968), pp. 101-35, 385-92; Brigadier General Lorenzo Thomas, Adjutant-General, U.S. Army, to Edwin M. Stanton, Secretary of War, October 24, 1863, *Official Records*, Ser. 3, Vol. 3: 916-17.

3. U.S. Congress, House, Testimony of Major General Lorenzo Thomas, April 17, 1866, Joint Committee on Reconstruction, 39th Cong., 1st Sess., 1866, *H. Rept.* 30, Pt. 4, pp. 140-44; *New York Times*, August 10, 1863, pp. 1-3.

4. First Lieutenant W. McGrew, Assistant Provost Marshal, Concordia Louisiana, to Brigadier General Lorenzo Thomas, June 14, 1864, Records of the Bureau of Refugees, Freedmen, and Abandoned Lands (hereinafter cited as BRFAL), Record Group 105, National Archives; Major General S. A. Hurbut to Lieutenant Colonel C. T. Christensen, January 3, 1865, *Official Records*, Ser. 1, Vol. 48: 401; circular published in Natchez, Mississippi, October 27, 1863, by Brigadier General Lorenzo Thomas, Adjutant-General, U.S. Army, *Official Records*, Ser. 3, Vol. 3: 939-40.

5. Major General Hurbut to Lieutenant Colonel Christensen, *Official Records*; *Natchez Democrat*, February 5, 1866, p. 2, February 12, 1866, p. 1; *Concordia Intelligencer*, July 26, 1867: 2; U.S. Congress, Senate, Letter of the Secretary of War Communicating a Copy of General A. A. Humphrey's Report to the War Department on the Levees of Mississippi, 40th Cong., 1st Sess., 1866 to accompany *S. Rept.* 126, p. 187.

6. See Thomas W. Knox, *Camp-Fire and Cotton-Field* (New York: Blelock & Co., 1865), pp. 305-440.

7. William Faulkner, *The Unvanquished* (New York: Random House, 1938), pp. 94-95, 116-117.

8. Ransom to Clark, July 16, 1863, *Official Records*, Ser. 1, Vol. 24: 680-81.

9. John Eaton, *Grant, Lincoln, and the Freedman* (New York: Longmans, Green & Co., 1907), pp. 1-18; Dorothy Lois Ellis, "The Transition from Slave Labor to Free Labor, with Special Reference to Louisiana" (M.A. thesis, Louisiana State University, 1932), pp. 1-14; Bell Irvin Wiley, *Southern Negroes, 1861-1865* (New Haven, Conn.: Yale University Press, 1938), pp. 184-90; Martha Mitchell Bigelow, "Freedmen of the Mississippi Valley, 1862-1865," *Civil War History* 8 (March 1962), pp. 38-47; Ulysses S. Grant, *Personal Memoirs of U. S. Grant*, 2 vols. (New York: C. L. Webster & Co.), 1: 424-26.

10. Eaton, *Grant, Lincoln, and the Freedmen*, pp. 1-18; Grant, *Personal Memoirs*, 1: 424-26.

11. Grant, *Personal Memoirs*, 1: 424-26.

12. Eaton, *Grant, Lincoln, and the Freedmen*, pp. 18-29.

13. General H. Halleck to General U. S. Grant, U.S. Army, March 31, 1-63, *Official Records*, Ser. 1, Vol. 24: 156-57.

14. Bigelow, "Freedmen of the Mississippi Valley," p. 43.

15. Eaton, *Grant, Lincoln, and the Freedmen*, pp. 46-61; Special Order of General Lorenzo Thomas, No. 63, September 29, 1863, BRFAL, Record Group 752.

16. See Brigadier General L. Thomas to Major General Charles A. Dana, January 26, 1865, Records of the Adjutant General's Office, Generals' Papers, Record Group 94; to Brigadier General M. M. Cuacker, Commander-District of Natchez, October 19, 1863, Records of the Adjutant General's Office, Generals' Papers (hereinafter cited as AGO-GP), Record Group 94, National Archives, Washington, D.C. It was common knowledge that Thomas went out of his way to aid those large Natchez District planters who had sworn their loyalty to the Union—many of whom were his close friends. On the other hand, he did not assist those planters who had supported the Confederacy until they had proven their loyalty.

17. BRFAL, Record Group 752; William C. Harris, *Presidential Reconstruction in Mississippi* (Baton Rouge, La.: Louisiana State University Press, 1967), p. 93.

18. Eaton, *Grant, Lincoln, and the Freedmen*, p. 59.

19. Ibid.; BRFAL, Record Group 752.

20. Eaton, *Grant, Lincoln, and the Freedmen*, pp 147-48; Knox, *Camp-Fire and Cotton-Field*, pp. 305-440; Major General R. S. Canby, U.S. Army, to E. Stanton, December 6, 1864, *Official Records*, Ser. 1, Vol. 48: 744-45; Preliminary Report on the condition and management of Emancipated Refugees: made to the Secretary of the War by the American Freedmen's Inquiry Commission (AFIC), June 30, 1863, in AGO-AFIC.

21. Brigadier General J. W. Davidson to Lieutenant Colonel C. T. Christensen, U.S. Army, Natchez, February 4, 1865, *Official Records*, Ser. 1, Vol. 48: 744-45; Report of Assistant Special Agent R. S. Hart to Special Agent William P. Mellon, August 6, 1864, Treasury Department, Civil War Special Agency Records, Record Group 366, National Archives, Washington, D.C.

22. George R. Bentley, *A History of the Freedmen's Bureau* (Reprint Edition, New York: Octagon Books, 1970), pp. 25-29; Bigelow, "Freedmen of the Mississippi Valley," pp. 38-47, Eaton, *Grant, Lincoln and the Freedmen*, pp. 142-66; Louis S. Gerteis, *From Contraband to Freedman: Federal Policy Towards Southern Blacks, 1861-1865* (Westport, Conn.: Greenwood Press, 1973), pp. 135-52.

23. Eaton, *Grant, Lincoln, and the Freedmen*, pp. 142-66.

24. Ibid.

25. L. Thomas to E. M. Stanton, February 20, 1864, *Official Records*, Ser. 3, Vol. 4: 24; March 14, 1864, *Official Records*, Ser. 3, Vol. 4: 176-77; L. Thomas to E. M Stanton, March 18, 1864, AGO-GP; L. Thomas to Major General W. T. Sherman, Commander, Military Division of Mississippi, *Official Records*, Ser. 3, Vol. 4: 210-11; L. Thomas to Lieutenant E. D. Townsend, Assistant Adjutant General, U.S. Army, Washington, D.C., April 19, 1864, *Official Records*, Ser. 3, Vol. 4: 225.

26. Special Order of General Lorenzo Thomas, No 9, March 11, 1864, *Official Records*, Ser. 3, Vol. 4: 166-70.

27. Ibid.

28. Ibid.

29. Ibid.

30. See Testimony of Brigadier General James S. Wadsworth before the Freedman's Inquiry Commission, AGO-AFIC.

31. Robert Dale Owen, James McKaye, Samuel G. Howe, Commissioner, the American Freedmen's Inquiry Commission, to E. M. Stanton, June 30, 1863, *Official Records*, Ser. 3, Vol. 3: 430-54.

32. R. S. Canby to E. M. Stanton, December 6, 1864, *Official Records*, Ser. 1, Vol. 48: 441-43.

33. See Steven Joseph Ross, "Freed Soil, Freed Labor, Freed Men: John Eaton and the Davis Bend Experiment," *Journal of Southern History* 44 (May 1978), pp. 213-32.

34. Bentley, *History of the Freedmen's Bureau*, pp. 49-102; William S. McFeely, *Yankee Stepfather: A Study of General O. O. Howard and the Freedmen's Bureau* (New Haven, Conn.: Yale University Press, 1968), pp. 64-83.

35. The point here is not that a program designed to make the former slaves viable landowners was a reasonable alternative at the time. Any such program would have floundered in view of the subsequent decline in the selling price for cotton and without a long-range subsidy or low-interest loan program. But the often mentioned alternative of self-sufficient, nonmarket-oriented landowners was a real possibility, although never clearly formulated by any of the involved leaders. The main dilemmas of such a program were those of property confiscation and the paternalism necessary to promote self-sufficiency in a market-oriented economy. With property confiscation ruled out, the concept of self-sufficient farming status for the freedmen lost all possible chance for success.

36. Eaton, *Grant, Lincoln, and the Freedmen*, pp. 16, 33, 106, 242-45; *Official Records*, Ser. 1, Vol. 48: 964.

37. Colonel S. Thomas to O. O. Howard, Commissioner, Bureau of Refugees, Freedmen, and Abandoned Lands, June 28, BRFAL, Record Group 105.

38. S. Thomas to O. O. Howard, June 27, 1865, BRFAL, Record Group 105; U.S. Congress, House, Circular of Lieutenant S. E. Eldridge, Assistant Commissioner, Bureau of Refugees, Freedmen, and Abandoned Lands for State of Mississippi, No. 7, July 29, 1865, 39th Cong., 1st Sess., 1865, *House Exec. Doc.* 70, pp. 154-56; Lieutenant G. R. Reynolds, Provost Marshal, Natchez District, to Lieutenant S. E. Eldridge, October 5, 1865, BRFAL, Record Group 105.

39. U.S. Congress, House, General Order Issued by Captain J. H. Weber, Acting Assistant Adjutant General, Bureau of Refugees, Freedmen, and Abandoned Lands. Office of the Assistant Commissioner for State of Mississippi, Order No. 13, October 31, 1865, 39th Cong., 1st Sess., 1865, *House Exec. Doc.* 70, pp. 265-71.

40. S. Thomas to O. O. Howard, June 27, 1865, BRFAL, Record Group 752.

41. U.S. Congress, House, General Order issued by Colonel S. Thomas, Assistant Commissioner, Bureau of Refugees, Freedmen, and Abandoned Lands for the District of Mississippi, Order No 16, December 31, 1865, 40th Cong., 1st Sess., 1865, published in *House Exec. Doc.* 70, p. 177.

42. U.S. Congress, House, Testimony of W.A.P. Dillingham, Treasury Agent at Natchez in the Fall of 1865, March 1, 1866, Joint Committee on

Reconstruction, 39th Cong., 1st Sess., 1866, *H. Rept.* 30, Pt. 3, pp. 116-19; S. Thomas to O. O. Howard, February 5, 1866, BRFAL, Record Group 752.

43. S. Thomas to O. O. Howard, February 5, 1866, BRFAL, Record Group 752.

44. Ibid.; S. E. Eldridge to O. O. Howard, October 10, 1865, BRFAL, Record Group 752.

45. S. Thomas to O. O. Howard, September 21, 1865, BRFAL, Record Group 752.

46. Colonel S. M. Preston, 58th U.S. Colored Infantry, Natchez, Mississippi, to Major J. S. Lewis, Office of the Assistant Commissioner, Bureau of Refugees, Freedmen, and Abandoned Lands for the State of Mississippi, December 17, 1865, BRFAL, Record Group 752; Captain J. Forgan, Sub-Commissioner, Bureau of Refugees, Freedmen, and Abandoned Lands for the District of Vidalia, Louisiana, to S. Thomas, January 31, 1866, BRFAL, Record Group 105.

47. Circular of Major G. D. Reynolds, Provost Marshal of Freedmen, 8th District, Natchez, Mississippi, June 1, 1865, BRFAL, Record Group 752.

48. G. D. Reynolds to S. E. Eldridge, February 2, 1866, BRFAL, Record Group 105; Captain J. H. West, Sub-Commissioner, Concordia Parish, Louisiana, to G. D. Reynolds, October 20, 1865, October 31, 1865, December 31, 1865, BRFAL, Record Group 105.

49. It should not be overlooked, however, that the bureau had organized several plantations in the vicinity of Vicksburg on a share tenant basis. A prominent Vicksburg merchant advanced supplies to the freedmen in return for a hold on the crops in what was undoubtedly an important precedent and examples of the possibilities available to freedmen and merchants alike. See U.S. Congress, House, S. Thomas to O. O. Howard, January 2, 1866, 39th Cong., 1st Sess., 1866, *House Exec. Doc.* 70, pp. 265-71.

50. *Natchez Democrat*, August 20, 1866, p. 3, August 31, 1866, p. 3.

51. G. D. Reynolds to S. Thomas, May 5, 1865, BRFAL, Record Group 105.

52. Ibid.; G. D. Reynolds to W. E. Strong, Inspector General, March 25, 1865; U.S. Congress, House, General Order Issued by Colonel S. Thomas, Order No 13, October 31, 1865, 39th Cong., 1st Sess., 1865, *House Exec. Doc.* 70, pp. 173-74.

53. *Natchez Democrat*, February 5, 1866, p. 2, February 26, 1866, p. 4.

54. G. D. Reynolds to S. Thomas, August 1, 1865, BRFAL, Record Group 105; *Natchez Democrat*, April 23, 1866, p. 1.

55. *Natchez Democrat*, February 5, 1866, p. 2, February 26, 1866, p. 4

56. G. D. Reynolds to W. Dix, Mayor of Natchez, March 6, 1866, BRFAL, Record Group 105.

57. G. D. Reynolds to General Carl Schurz, September 26, 1865, BRFAL, Record Group 105. Reynolds explained to Schurz that the freedmen in the Natchez District would probably remain on "their own places this year . . . but next year they will leave and make other arrangements."

58. Ibid.; G. R. Reynolds to S. E. Eldridge, October 5, 1865, BRFAL, Record Group 105; Dillingham, Joint Committee on Reconstruction, 39th Cong., 1st Sess., 1866, *H. Rept.* 30.

59. See Bentley, *History of the Freedmen's Bureau*, p. 133; McFeely, *Yankee Stepfather*, pp. 70-71.

60. S. E. Eldridge to General T. J. Wood, U.S. Army, May 11, 1866, BRFAL, Record Group 105; see also Bentley, *History of the Freedmen's Bureau*, pp. 215, 244.

61. James W. Garner, *Reconstruction in Mississippi* (Baton Rouge, La.: Louisiana State University Press, 1901, 1968), pp. 107-260; James E. Sefton, *The United States Army and Reconstruction, 1865-1877* (Baton Rouge, La.: Louisiana State University Press, 1967), pp. 46-49.

62. Bentley, *History of the Freedmen's Bureau*, pp. 198-99; Garner, *Reconstruction in Mississippi*, pp. 184-85; Sefton, *United States Army and Reconstruction*, pp. 64, 89-90.

63. Garner, *Reconstruction in Mississippi*, p. 266; Harris, *Presidential Reconstruction in Mississippi*, p. 77.

64. Garner, *Reconstruction in Mississippi*, p. 107.

65. Bentley, *History of the Freedmen's Bureau*, pp. 198-99.

66. U.S. Congress, Senate, T. J. Wood to O. O. Howard, October 31, 1866, 39th Cong., 2nd Sess., 1866, *Senate Exec. Doc.* 6, pp. 94-100.

67. T. J. Wood to O. O. Howard, December 10, 1866, BRFAL, Record Group 752.

68. Ibid., August 9, 1866, October 12, 1866, December 10, 1866.

69. Ibid., October 12, 1866.

70. Ibid., December 10, 1866.

71. Ibid., October 19, 1866.

72. Ibid., December 10, 1866.

73. See L. Thomas to W. P. Mellon, February 22, 1865, AGO-AFIC. Thomas believed that any type of established regulation wherein the army would be forced to intervene on behalf of the freedmen would ultimately work to their disadvantage by making them dependent upon the government. In his mind the planter should determine wages due the freedmen by a judgment of the work that was done.

• 4 •

"...ALL OF THE NEGROES
WANT TO WORK NEXT YEAR
ON SHARES"

Although the U.S. Army generally favored fixed wages as the ideal remuneration for blacks in the Natchez District, relatively few of the neighborhood's freedmen worked as wage laborers a generation after the Emancipation Proclamation. Rather, the vast majority were share-croppers and tenants. If we judge from the manuscript census data on status and family, between thirteen thousand and fifteen thousand freedmen fell into this category in 1880, and the large majority (some eleven thousand) were sharecroppers and their dependents. As for the rest, if we look at the wage schedules (listed in the unpublished manuscript census) of those farmers who paid wages for more than sixteen weeks in the year, it can be estimated that approximately one thousand freedmen were employed as full-time, fixed wage hands. The remaining amounts listed in the wages columns were monies paid to employ pickers during the harvest season. The average picker worked from four to six weeks for wages ranging from 40 to 60 cents a day as part of a marginal labor force that numbered around twenty-seven hundred people—men, women, and children. Although little is known about the status of these marginal hands during the rest of the year, it is clear that sharecropping had become the dominant form of land tenure and employment for freedmen in the neighborhood by 1880.[1] (See Appendix C.)

But as mentioned earlier sharecropping did not emerge overnight. Only a few planters employed the arrangement in 1866, and in no instance was the share system anything but a method of wage settlement in which the freedmen worked in gangs, or squads, as hired hands. They received wages plus a small share of the crops, usually one-tenth, or a small share of the farm's net proceeds for their pay.[2] The system did not differ at all from the more common wage system in the actual conditions of labor: gang labor and rations and strict supervision prevailed in all cases.

A careful study of the mortgage records and papers of the Freedmen's Bureau indicates that the typical planter tried working his former slaves on this combined fixed wage and small share basis in 1864. He then moved to shares, usually one-third to one-half of the crops as pay, and to share tenancy within three or four years after the war, or to that system whereby the freedmen paid one-half to two-thirds of the crop as rent. By 1880, a number of landlords were dealing with their once hired hands on a fixed rent basis. Some form of sharecropping or fixed rent tenancy, with landlords and freedmen shifting between these forms, became the established labor arrangement in the district.[3]

Some of the clearest insights into the early stages of this development are found in the plantation documents of William Mercer, a Natchez District planter who achieved the status of planter elite several decades before the war and thereafter lived the life of a refined gentleman in the best of the worlds available to his class. During his long stays in New Orleans, Mercer's three plantations were managed by one of the most industrious and loyal overseers for whom historians have evidence, William B. Shields. As much a personal friend as an employee, Shields had worked for twenty years on Mercer's behalf when the Civil War so dramatically changed his life and enterprise.[4]

Although Shields continued in Mercer's employ after the war's end, the new situation strained the old manager's experience and ingenuity to the limits. Taxes were high, the levees ruined, and politics frustrating and uncertain. But most disruptive and difficult was the fact that blacks were now free people with labor to sell. For the first time in his life, Shields found himself bargaining with labor rather than commanding it, haggling rather than giving undisputed orders.

Most of Mercer's former slaves had worked for their old master in 1864 for fixed wages, then for a share of the cotton in 1865, and finally for a combination of fixed and share wages in 1866. The last-named situation found them working for $12 a month and a one-tenth share of the cotton made. Although the details are not clear, the freedmen themselves seemingly initiated the contract variations in hopes of increasing their incomes. There is no evidence that the move to share wages involved a contribution by the hands to a part of the equipment and supplies needed in production. The low wage rates set by the army in 1864 had motivated the freedmen to demand wages in shares once they were free to negotiate, and neither Shields nor Mercer objected to the alterations.[5]

But as 1866 drew to a close, Shields urgently informed Mercer that the freedmen had refused to sign contracts and were demanding Saturdays off, horses and mules to ride to Natchez, and more pay.[6] The old "Nabob's" reaction was immediate, stern, and tactful: although the amount and form of pay increases were negotiable, under no circumstance would Shields be allowed to give the hands time or animals.[7] Instead, Mercer instructed his manager to offer double the present cash wages, while dropping the one-tenth share, and to do what he could to weed out the troublemakers.[8]

But the old planter's decision failed to settle the matter. Unhappy with their terms of contract, the hands became increasingly insubordinate. No longer, for instance, could Shields expect to buy needed poultry and eggs from the freedmen at bargain rates as they raised prices and began selling to the highest bidders.[9] Infuriating, too, was the haughtiness and pretension with which the freedmen carried out their tasks. The women according to Shields were practically out of control: "As soon as one of them conceives or thinks herself pregnant, she gives up work altogether.[10]

Especially vexing were the efforts of labor to escape from the manager's watchful eye. More than any other remnant of bondage, the closely supervised gang-working arrangements struck the emancipated slaves as too confining to tolerate. Working from sunup to sunset in clearing, plowing, and weeding gangs enabled the white manager and landlord to exercise a control over the freedman's daily existence that included physical punishment in some cases and constant overlording as a general rule. As long as white bosses directed field gangs, blacks felt like slaves. The sentiment was apparently universal and finally found expression among Mercer's hands when, in late 1866, those at his Armonde plantation demanded to work in family squads under their own supervision.[11] Although the request said nothing of tenancy, Mercer thought of the proposal as the first step towards an unacceptable independence for his freed slaves. So dismayed was the old man at this turn of events, that he thereafter refused to grant shares as wages even for closely supervised gang labor and vowed to give up planting before adopting a system that would undermine his absolute authority over his hands.[12]

What had enabled freemen even to challenge the authority of a "Nabob" like Mercer? The Freedmen's Bureau, although preferring fixed wages, stayed out of such disputes as much as possible in the hope that bargaining would be good experience for both parties. Other factors, however, were at work. "Our immediate neighbors are the cause of all this doubting," explained Shields in a revealing note to Mercer in December 1866:

> Metcalf I hear is making efforts to get a very large force, offering inducements with plenty of whiskey and every latitude and liberty to do as they please, so they work for him. And Hutchins tampers with our Negroes and those who left us, particularly old Daniel at E, offering to furnish mules, utensils, and all plantation gear and tools for half the cotton made. . . . and if Hutchins succeeds in making this arrangement with any number of those to whom his overtures have been made, we may bid farewell to all our quadruped—at least of the hog and many of the cattle—for his Negroes, as on other neighboring places, are to furnish their own meal and bread.[13]

And the "neighbors" did succeed. By the end of the year, the Buckhert plantation was nearly deserted while the Armond place lost eighteen hands and Elliscliff five.[14] Hence, the competition for labor by would-be employers on every side had given freedmen the opportunity to demand unprecedented concessions.

Mercer remained true to his word, however, and refused to hire anyone but closely supervised fixed wage laborers. By 1870, his once substantial labor force reduced to a handful, the old planter began leasing portions of his holdings to whites who most likely contracted with freedmen as share laborers and tenants to grow cotton.[15]

Mercer's experience with the freedmen was not an isolated case but the normal pattern in the Natchez District. Planter Lemuel Parker Conner, owner of several district plantations and a well-established member of the neighborhood's plantation elite, had even less success in controlling the wage arrangements with his labor.[16] The first indication of trouble came while he was away at war serving as an army officer for the Confederacy. From home arrived word of a plot among fourteen of his slaves, one of whom had betrayed its leader exposing the plan in time to prevent it. Apparently, the scheme involved little more than talk among a group of slaves about their plans to kill their masters once northern soldiers appeared in the Natchez District.[17] In 1863, Mercer's wife wrote worried tales to Conner describing the restlessness of slaves on their Concordia plantation, Innesfield. She reported that several slaves were so bold as to openly hail a Yankee gunboat to make good their escape.[18] At the war's end, Conner vigorously worked to resume planting on the family's Home Place, Innesfield, Killarny, and Rifle Point plantations. Conner's contracts with the freedmen in 1865 included a share settlement as an alternative to fixed wages, leaving the choice up to the freedmen. The terms called for half of the wages to be paid monthly, with the remainder paid at the end of the year, or a one-tenth share of the net profits in lieu of the wages due.[19] It is not known what motivated Conner to make this offer, except that his leaving the choice to the workers suggests that he was not looking to share the risks of production with his labor. At the most, he probably thought it a good incentive for promoting industrious labor habits among his former slaves. It is more likely that share wages were terms demanded by his black workers.

Conner's brother, Richard, often complained of his inability to find workers except under some share arrangement. In August 1866, the brother noted that "all of the Negroes want to work next year on shares." The most that he felt could be done about these demands was to accept them quickly before his best hands contracted elsewhere. He suggested that if the brothers contracted for one-half the crops made, "we supplying farming implements and they paying for rations and for any extra labor," they could yet "select those Negroes who have been truest to their contracts and worked during this last year."[20] One year later, Conner's Innesfield plantation in Concordia erupted with near revolution as the workers refused to "give up the cotton and settle for the half wages due." Although the details are not clear, it seems that the hands wished to keep a share of the cotton rather than accept either the wages or a share in the "net profits." What is clear is that the blacks were in a position to keep the cotton already gathered, they were determined not to pick more, and they were willing to make a stand. Similar agitation prevailed at his Killarny plantation. In the end, Conner "brought a Bureau agent down to help him persuade the blacks at Innesfield," and "settled the matter very easily at Killarny by discharging sixteen hands."[21]

This kind of settlement did not last long, however. From that time to his death in 1891, neither Conner nor his brother nor anyone in the

Conner family found a solution to the "Negro problem," which they often described as the burden of attempting to work with "faithless creatures."[22] As late as 1880, Conner's brother still complained that each year "the best men" on his place threatened to leave "unless allowed to rent."[23]

A close reading of these plantation documents reveals several important points about the origins of sharecropping. In the first place, there is little evidence that the former slaves' preference for shares was based on a clear understanding of what the system might become. In general, freedmen believed that shares, as an alternative to the low wages established by the army in 1864, offered the best chance of securing the higher pay compatible with the selling price of cotton.[24] In addition, although there is no evidence that Shields or Conner had cheated hands of wages, many neighboring freedmen had suffered such treatment so few blacks were unaware of the problems posed by this form of wage system.

From the planter's view, as suggested by Mercer's early flexibility, most well understood that share wages little affected their interests. In the case of Mercer, the old planter's mind was open on the issue as long as he retained complete supervision of the blacks. He must have realized that in either setting the freedman's earnings would be largely consumed through the plantation store long before the harvest. Equally important, the planter's control over the freedman's working day and wage settlement meant that wages were subject to deductions for time lost, insubordination, inefficient labor, and numerous other items of conduct.[25]

Freedmen soon learned that such shares neither safeguarded them from ending the year in debt to the store nor provided sufficient income for the purchase of land. This explains why blacks throughout the district changed the terms of their contracts so frequently. Freedmen jumped from shares to fixed wages and back again in what amounted to a desperate struggle to solve the problem. They began demanding concessions unrelated at first glance to wages, once the extent of their no-capital no-credit dilemma was fully realized. Unable to end the year solvent, the freedmen saw independence as a key bargaining concern. The right to leave the plantation in the evening at will, to take their guns to the fields for a little mid-day hunting, to work a patch for growing vegetables, to have poultry and pigs, to own title to a portion of the crop to facilitate their control over its disposal, and to contract and work in family squads or groups were upmost in the minds of Natchez District freedmen by 1869.[26] Such conditions of contract meant freedom in an everyday sense and generally enabled the most industrious to increase their real incomes as well. Those who won such terms were free to travel from plantation to plantation at day's end for what planters believed was an endless series of parties, but also certainly to sell their truck, catch an occasional opossum for the dinner table, and visit relations and sweethearts.[27]

Most importantly, freedmen wanted shares in order to escape the planter's ability to discipline them by means of wage forfeiture. The freedmen's experience with the wage system employed since 1863 so strongly identified wages with slavery that there was no possibility that district blacks would ever view the system as a fair means of labor compensation or as a desirable work setting for free men. Not only were the fixed wages abysmally low, consumed through the plantation store, and often nonexistent, but also they were always used as a means of forcing freedmen to labor according to the landlord's dictates. It is no exaggeration to say that wages, as the U.S. Army had hoped, replaced the whip as the chief form of labor compulsion in 1865, with the former slave's every action subject to the threat of wage penalties. The records of the Freedmen's Bureau are vivid testimony to how quickly landlords learned to safeguard their profits by means of wage deductions. A few examples of the dozens available should suffice. W. Lowenberg, lessee of Tacany plantation in Concordia Parish employed sixty-seven freedmen in 1865 from January to December. Wages ranged from $4 to $15 per month, with most paid $7 to $10. Lowenberg's total wage bill for April amounted to $552, of which $145.65, or over 26 percent of the wages due, was deducted for conduct ranging from starting late in the morning, failing to be in the field by the agreed time after lunch, unexcused absences, or sickness. The average loss was $2.17 per hand, although seventeen hands lost more than the average that month.[28]

Another planter, William H. Wheeler, lessee of Sycamore plantation, employed eighty-one hands in 1865. His total wage bill in March came to $653.00, or an average wage of $8.06 per month per worker, from which $180.88, or an average of $2.23 per hand, was deducted for wages lost. In this case, the records show the total days each worker lost as the basis for the wage deduction. The entire work force lost a total of

671 days, or an average of 8 1/4 days per worker. Two people lost twenty-eight days each, and only six lost less than seven days. If we assume that these rates of loss were typical, and planter testimony suggests they were, Wheeler deducted $2,170.56 from his yearly wage bill of $7,836 and his workers lost a total of 8,052 days, or an average of 99 1/2 days each. This meant that as a result of wage deductions, each worker lost $26.80 from his total earning of $96.74, or more than one-quarter of his wages.[29]

Each freedman did not stay away from the job the number of days cited as "lost." Rather, these days were the total accumulation of deductions for insubordination, being late, "idle labor," and other less tangible claims. Hands likely worked the majority of days for which the deductions were made, but did so in ways that were not satisfying to the planter. The "remarks" of lessee J. F. Evans of Glen Aubin plantation in 1865 were typical:

I have further to report that during the last month the laborers of this plantation have been doing short work. With the exception of three, none of the laborers go out to work at the appointed hour either morning or afternoon. When in the field the majority of them actually idle away their time. Compared with the work done on this plantation under a compulsory system, three laborers do now hardly as much as one did then. Besides idling away their time in the field some feign illness and lay up for whole days, others absent themselves for the greater portion of the day attending their own affairs.[30]

Similar comments are scattered heavily throughout the records of the district, but few were so precise. Most simply noted, "all work very lazily done—do half as much as they ought to," or remarks such as "indolent" next to the freedman's name on the reports.[31]

Few planters took the time to explain fully to their hands what they expected of them, but few freedmen ended 1865 with any doubts of the power their employers held in the matter of wage forfeiture. In most contracts, the freedmen were paid half the wages due each month minus the full deductions or wage penalty of that month. In the examples used above, this meant that the average hand on the Sycamore plantation received $2.92 in wages each month, with the remainder to be paid in a lump sum at the year's end—that is, if any wage was left once the freedman's bill for purchases made on credit in the plantation store was deducted from his remaining wages. In these circumstances, it was not

uncommon for freedmen to end the year in debt to the store, with no
wages coming, and hopeful of a small cash advance against next year's
wage. Regardless of the individual case, most freedmen understood that
any wage system that enabled planters to discipline their conduct by
wage forfeiture seriously undermined both their newfound right not to
work likes slaves and their right to fair pay for fair labor—which in their
minds meant pay beyond subsistence rations.

The freedman's ability to sell his labor depended, in turn, upon the
area planters' willingness to offer terms that Shields considered "even
too absurd to mention." Some leased entire plantations to groups of
freedmen for a share of the crop, dealt only with families, or negotiated
with individual freedmen who then hired laborers on their own.[32] The
situation was seldom contingent upon the presence of northern soldiers
on southern soil: planters needed hands to work fields that promised
high profits. The going price for middling cotton at New Orleans, 42
cents a pound, not only enticed northern capitalists to lease lands and
compete for labor through contract variations but also drove friends of
long standing to contest one another in unheard of fashion. Shields
regretfully noted in one telling report to Mercer that

our neighbor Pipes has also entered the list against us, and like the others had
induced some to go with him. There can be no question of the fact that Metcalf
has been tampering with our Negroes—cajoling, and urging them to leave. I
only harp on these things to show you what some of our neighbors have come
to, and that you may see how temptation is thrown in the way of the Negroes.[33]

Profits from the furnishing business led many planters to accept the
freedman's demand for shares. It was not that sharecropping better
enabled planters and merchants to secure the supply business of the
freedmen, but that the system offered no impediment to the doing of this
business. Another of Mercer's neighbors, John Hutchins, for instance,
in addition to supplying food and clothes to the freedmen through his
plantation store, did such a good business in selling whiskey that Shields
felt he was doing "better at the present time than he did under the old
regime."[34]

The sources also indicate that the low labor supply forced planters in
the neighborhood to compete vigorously with one another. Professional
employment agencies operating out of New Orleans did a brisk business

supplying the district with field hands on a commission basis, while the Freedmen's Bureau did what it could to transport blacks from the interior to the delta and swamps.[35] Labor was so scarce because many blacks had died during the war, not only in battle but also from sickness and exposure. The literature abounds with the murders of plantation blacks by Confederate guerrillas, while others lost their lives as Union soldiers.[36] In addition, the war relocated thousands, and more than a few never returned to their slave neighborhoods. Still others left the area immediately after the war to rejoin families broken up by slavery.[37]

But the major source of labor scarcity flowed from the freedman's reluctance to contract. Thousands held off contracting in 1865 in the hope that the army would soon distribute lands. Countless others, refusing to return to the swamps and lowlands of Concordia Parish, flocked to Natchez or lived off the land as best they could. Entire villages of freedmen survived for years by cutting and selling firewood to the river boats.[38] Of those who refused to contract, most did so mainly because they recognized the strength of their position. Unless the district's planters reached some agreement among themselves about wages and terms of contract, nothing prevented the freedmen from selling their labor to the highest bidder. One old farmer writing in 1873 to a neighbor recalled how district planters had brought shares on themselves by refusing to set wages in the immediate postwar period: "Each one wished to be unrestrained and free to act as judgment might direct. The result is before us. The time is now approaching, if not now, when a premium will be asked to have your lands occupied."[39] Conner's brother, another long-standing member of the district's plantation elite, echoed these sentiments exactly: "The devil with these Negroes, and it must always be the case until the planters in every county get together and establish some maximum figure to give freedmen. And for a violation the person should be driven from the country."[40] As long as planters vied with one another for labor, freedmen were able to hold themselves off the market. Even when the contracts were signed, the planter had little guarantee that his hands would stay through the season. As a result, planters tended to overcontract, thus putting additional pressure on the labor supply. Shields described one particularly aggressive northern capitalist who vowed to have as many freedmen on his place as possible to counter the freedman's tendency to desert regardless of contracts signed:

This year he gave twelve dollars and allowed them (freedmen) to plant as much cotton as they pleased, whiskey and stealing not counted—and as many on the place as chose to come, he only paying and provisioning those who work—thus he has had some two hundred hands staying there during the year, his ration list averaging about sixty hands—some of the two hundred working one day—some another. He clears much from the rum shop—will make between 300 and 400 bales and the Negroes make 40 or 50 bales besides their pay. For the coming year he proposed to extend this system and is offering greater inducements hence the cry for the swamps. I hear they are now picking, in the swamp, 1,000 to 12,000 to the acre.[41]

SHARECROPPING VERSUS WAGES

Let us now examine why the freedman's demand for independence and income, and the planter's demand for labor, resulted in sharecropping rather than wages. Surely, nothing in the nature of wages prevented planters from working their hands in family arrangements, giving cabins and gardens, and allowing them the freedom to do as they wished on Sundays. Nor would the profits from supplying fixed wage hands have been any less than those made from sharecropping. Wage laborers have traditionally found the opportunity of earning pay and bonuses incentive enough to work long and hard and well, even with minimum supervision.

One possibility is that planters introduced sharecropping as an incentive for freedmen to increase their personal efficiency by giving them a substantial interest in the crops. But if Mercer's "neighbors" were typical, there is little basis for suspecting that planters thought of sharecropping, at least in the beginning, as a means of increasing the freedman's personal efficiency. Those planters who adopted shares as a method of wage settlement for gang labor seemingly did so at the freedman's insistence, while those who utilized the system as a form of tenancy showed little concern with motivating the freedman to greater levels of productivity. They were interested primarily in securing laborers to work the land and make a crop.

This is not to say that planters overlooked the advantages of giving freedmen an interest in the crop. Planter Samuel Postlethwaite justified his move towards shares because the system better enabled him to keep the freedmen working through picking time.[42] Although it is unclear why shares held labor for the entire season in comparison to wages

payable when the crops were sold, his theory undoubtedly had some basis in reality. Perhaps the blacks felt that sharecropping was a better protection of earnings, since they could "see" their "property" and "know" its value, at least in comparison to a fixed wage system in which their earnings were subject to deductions and forfeiture for misconduct. But this only suggests that a fixed wage system, rather than share wages, may have worked if administered with an eye to satisfying the needs of the freedmen as former slaves anxious to earn capital for buying farms or other investments.[43]

Another possible explanation of the move to shares is that the system lowered the managerial costs of production as planters abandoned the practice of close supervision over the wage laborers. This possibility seems likely, since sharecropping normally involved family patches and less than rigorous supervision. But it would be misleading to assume too strong a casual relationship between the costs of supervision and the origins of sharecropping. It must be understood that few antebellum planters believed that blacks were capable of working long and hard and well with little supervision. From their perspective, the freedmen, like the slaves they had been, needed strict supervision in gang labor or at group tasks. The planter's goal after the war was to employ the freedmen at low wages under close supervision, using the system of wage penalties to insure labor discipline. At first, they thought only in terms of low fixed wages, as the U.S. Army had. Then, as freedmen felt cheated and were reluctant to work, except on shares, planters agreed to offer a small share of the crop along with low wages as an incentive for labor to contract. This share system abandoned neither gang labor nor strict supervision as such, although many northern lessees did agree to leave the freedmen to themselves in the fields. It was essentially a system of subsistence share wages. As in the case of Mercer, few planters objected to granting freedmen a small share of the crop as wages as long as they retained supervision over them. Had planters believed that freedmen would labor faithfully on their own, they could have abandoned low wages and supervision in favor of high wages and self-directed family or squad working arrangements, especially in the early days when cotton prices were high. But from their perspective, blacks needed supervision regardless of any other incentives provided. The issue was one of social control and the labor discipline of a people who were believed to be inherently inferior or so inexperienced as to be incapable of self-direction.

Landlords would not move to any form of unsupervised status for black workers if they could prevent it, regardless of the theoretical savings involved. It never occurred to them that the freedmen might respond to incentives other than supervision. They could not view blacks as capable of working independently, given meaningful bonuses or a share of the profits in place of supervision.

Because of this perspective, the old southern planter group came into conflict with with those few northern lessees and bureau agents who were convinced that freedmen would work well if paid well. It also separated them from those speculators (northern and southern) in the first few years after the war who were willing to try any labor arrangement to get the abandoned cotton picked or the fields planted.

Nor did the origins of sharecropping reflect the planter's desire to use the system as a means of forcing the freedmen to share in the risks of production, although this possibility seems likely in view of the natural and manmade calamities that plagued the lower Mississippi River Valley in the immediate postwar era. One typical situation is described in a letter to *The Cultivator and Country Gentleman* in 1868:

My cotton plantation is again a failure, and as this is the third year that it has proved so, I do not think that I shall try it again. The first year, as you are aware, my plantation was destroyed my stock and Negroes run off. Last year the army worms destroyed my cotton, and this year by the same pest, only more numerous. My plantation is at Vidalia, Louisiana, opposite Natchez. This year I cultivated 1,000 acres in cotton. By extraordinary exertions, I kept out of the water. I had a good stand on most of it, and it promised well until August, when the worms made appearance, and so thoroughly did they ravage it, that I shall not have 100 bales.[44]

In addition to such natural calamities, planters found the political situation frustrating and uncertain. Taxes on cotton were high in the immediate postwar period, and few planters knew what to make of the future. For the struggling planter attempting to get back on his feet, fixed wages undoubtedly posed a great risk should the crop fail for whatever reason.

Yet, the fact that most planters moved to sharecropping reluctantly suggests that few saw advantages to it in these terms, especially in view of the low fixed wages normally consumed in supplies or forfeited in wage deductions by year's end. Conner thought "giving the wretches an

interest in the crop a poor way to make money," but finally resigned himself to do so or abandon planting, since he could find few workers under any other system.[45] Even Stephen Duncan, Jr., heir to one of the largest fortunes in the district and a Union loyalist, told of the "disgusting business" which the freedmen had forced upon him—namely, sharecropping. As a result of his refusal to go along, Duncan lost over half the hands at his Oakley, Hollyridge, and Duncanon plantations.[46]

It is not that planters ultimately failed to see the advantages that sharecropping provided them in sharing the risks of production with their labor. They did, once they had no choice in the matter. Conner himself admitted he was glad he had contracted on shares in view of his losses. And once the system emerged, planters increasingly accepted it because of these advantages. Seeing the planters' theoretical advantages must not mislead us into thinking that they had wanted it in the first place. Moreover, their doubts about the freedmen's ability to work as self-directed laborers found them disregarding the use of share incentives as a substitute for control. Instead, planters struggled to retain enough authority in sharecropping to reduce the risks of working unsupervised blacks. Forced to give up gang labor, their resident status, and their general dictatorship over blacks, planters worked to insure their control over the crucial decisions of crop-making. Most planters never believed that shares enabled them to diffuse the risks of the freedman's supposed unsupervised inefficiency, and they did not believe blacks were capable of responding to shares as an incentive. Hence, they found sharecropping acceptable only if they could make the crucial decisions about what to plant, where to plant, and when to plant.

The point is clearly illustrated in the case of Mercer's neighbor, J. A. Gillespie, who worked his Hollywood plantation in a manner that was typical for the area.[47] The plantation was located about 3 miles due east of Mercer's Armonda plantation. Gillespie had abandoned fixed wages, mass contracts, and gang working conditions by 1868. In their place, the old planter contracted with squads usually centering around a family unit, or simply with people who got along together fairly well, paying them a share of the crop as wages. One typical squad, including eight men and three women, worked for one-third of the crop, with Gillespie paying half the cost of preparing the crop for market. Others on the place were hired as wage hands. Beyond the fact that the wage hands worked for a fixed wage, little separated them from the share hands in

the actual conditions of labor. Both groups worked in squads under contracts that stipulated the exact crops to be planted, the proportions, and the conditions under which the freedmen might obtain tools, teams, and supplies at a charge to be deducted from their wages or shares. In both instances, Gillespie retained overall supervisory capacity on the place, which meant ensuring the proper care of tools, teams, and property as well as the right to oversee the progress of the crops. If the crop was not progressing to his liking, Gillespie could even rewrite the contracts in midseason by employing additional hands at the squad's expense. Ginning equipment was available to the croppers, but only under the strictest supervision. All freedmen contracted to cut wood and shell corn for the Gillespie household.[48]

In 1869, Gillespie began contracting with his squads for one-half the crop as rent; the squads paid their expenses out of their share of the crop. But little had actually changed. Except that the fixed wage hands now functioned as a squad of extra hands to be employed at the expense of the tenants should they be needed to keep the crops in order, and that the share renters now paid Gillespie one-half the crops made as rent, all other stipulations in the contracts were essentially the same as before. Gillespie continued to supply rations and to decide what was to be planted, when, and where. The share tenants, too, continued to work as before under Gillespie's general supervision, and provided wood and corn for the old planter's family.[49]

The next year, in 1870, Gillespie began leasing lands to his former slaves for a fixed rent paid in cotton, while retaining his general control over the working conditions on the place. This practice remained Gillespie's set procedure for the next decade in a pattern duplicated throughout the district.[50] Freedmen on the Overton and Siego tracts in Adams County, for instance, worked in squads and in families and groups and as fixed rent tenants, with the right in each case to name their own foreman, punish troublemakers, and own stock. Nevertheless, they were obligated to obey the landlord's agent as to the form and methods of cultivation.[51] And it mattered little whether the freedmen worked as fixed renters or share renters. Another planter, S. G. Kennedy, contracted with freedmen as share renters, dividing with them the crop. Yet, the actual working conditions left Kennedy the right to class hands for the payment of a pro-rata share of their half of the crop, to hire other hands if necessary, with all wages paid at the expense of the sharecroppers, and to receive ten hours of labor per day from each cropper contracting.[52] Merchant Andrew Smart held similar authority over his fixed rent tenants in 1880. In his accounts, Smart leased 44 acres to two freedmen families for a rent of 9 1/2 bales of cotton. These tenants were not independent farmers left to their own motivations for work: "all cultivation, picking, ginning, etc.," according to the contract terms, "shall be done under the direction of Smart's manager with Smart deciding all differences." The landlord reserved the right to halt rations for neglect of the crop, and to police the use of mules, tools, and visitors to the place.[53]

Convinced that freedmen were incapable of laboring on their own, landlords managed to write contracts wherein they controlled the labor conditions regardless of the system. But it must be understood that although no real difference existed between share tenancy and fixed rent tenancy regarding the planters' ability to control the cultivation of the crops, there were major differences between sharecropping and the fixed wage setting of the immediate postwar years. All one need do to see the point is to read the contracts. Planter W. J. Minor's rules for working his Waterloo plantation in 1864 with freedmen read like those for working the place under slavery: (1) hands were required to work ten hours a day under gang labor conditions, (2) all lost time was deducted from their wages, (3) rations would be stopped if wage forfeiture failed to induce faithful labor, (4) no more than 50 cents a month would be paid at year's end to any hand owing work, (5) time going to and from work was not counted as part of the work day, (6) those working less than ten hours were docked 1/4 of a day, (7) insubordination meant 1/4 of a day to a week docked, (8) married women not working were to be supplied at the husband's expense, (9) nonworking wives and mothers and sisters were to be charged for board, room, and fuel, (10) the first morning bell sounded at 5:00 A.M., the second at 6:30 A.M., and the third at 7:00 A.M., at which time all hands had to be at work or docked 1/4 day. The noon bell sounded at 12:00 for lunch and at 1:30 P.M., at which time hands had to be at work or be docked 1/4 day, (11) none but regular teamsters were to use a team, and (12) no visitors were allowed on the place without papers and then only on Saturdays and Sundays. Minor kept these rules in force for as long as possible, with obvious effect on his labor supply. In 1866, he noted that his hands were refusing to contract, with "most leaving for Natchez or to work on other plantations."[54]

Compare these conditions with those which Martha Claiborne adopted in 1870 in dealing with her freedmen sharecroppers on the Dumbarton plantation. She stipulated that (1) sharecropper C. Carroll was to cultivate 80 acres at her expense for one-third of the crop raised, (2) her cotton was to be divided at the gin after being baled and ginned by Carroll—"she having use of the gin for that purpose," (3) she was to furnish her own teams and labor and pay for bagging and ties, (4) all other crops were to be divided at the field with Carroll present, and (5) Carroll could have her lease renewed if she conducted herself to Claiborne's satisfaction. Noticeably absent were specific rules governing Carroll's work conduct, daily routine, or system of wage penalties, although there can be little doubt that Claiborne was ready and able to bring in hired hands to aid the crop at Carroll's expense should there be the need.[55]

Even more revealing of the change were the contracts of Thomas R. Shields with his tenants on the Hermitage plantation in Adams County in 1870. Shields leased the entire place to twenty-six freedmen for five years for a rent of 18 bales of cotton per year. The freedmen worked in squads with the understanding that no matter what the total crop production was they would be responsible for no more than the 18 bales rent. In addition, (1) Shields agreed to furnish bagging, ties, twine, and the like, for the 18 bales, at his expense, (2) the plantation gin was to be left in good working order at year's end, (3) lessees were to work plantations at their own expense, (4) oxen, tools, wagons, and other plantation equipment on the plantation were available for all to use, but only for legitimate plantation business, (5) two men were assigned to function as carpenters and blacksmiths and as cattle tenders, with use of specified rooms in the plantation house, (6) tenants would pay fines for the destruction of property, (7) and a head man was named with specific instructions "not to interfere in individual operations except to prevent abuse of oxen or other things of common use."[56]

Under sharecroping as it functioned in the 1870s and 1880s in the Natchez District, freedmen won the right to work in groups and families, to labor generally under their own supervision in their daily labor, and to labor free of the wage penalty system of the immediate postwar era. These concessions were won by the freedman's resistance to labor practices similar to slavery, and no one involved considered them small victories.

In their clamor for share wages and share renting, freedmen wanted more than just an escape from close supervision. Had they received wages sufficient for them to provide their families in something more than subsistence style, they might have accepted labor conditions similar to Minor's terms, until they could buy land or pay a fixed rent. Their demands for independence were seldom unrelated to their interest in real income gains. The best evidence for this notion is in the fact that planters often found their "best hands" abandoning them or willing to stay depending upon the terms of contract. In the immediate postwar years, freedmen found employers willing to give them land for gardens, supplies, and even teams and tools to get them working. Some paid a share of the crop as a better risk than fixed wages. This precedent identified real income gains with sharecropping and soon fastened the system upon the Natchez District as the only terms for which the "best hands" worked. In time, because supervision costs were probably less with sharecropping relative to fixed wages, planters and landlords grew to tolerate the system as an acceptable substitute for one of low, fixed wages. But this was the end result of a generation of change and is not pertinent to an understanding of the origins of sharecropping in the Natchez District.

TENANCY VERSUS SHARECROPPING

Theoretically, freedmen should have preferred fixed rent tenancy over sharecropping, since everything produced over the rent belonged to the tenant alone. In addition, in view of the intense competition for labor, it is surprising that blacks failed to bargain for fixed rents.

Perhaps fixed rent tenancy was unacceptable to southern planters and landlords because they were convinced that blacks could not function as truly independent farmers, or possibly because they feared the rise of an independent class of black farmers. But the evidence suggests that neither fixed rent nor share tenancy differed much in the extent of supervision possible. Regardless of the contract, the landlord's control over the crop enabled him to obviate any supposed variation in risk between the two. In fixed rent tenancy as in sharecropping, the landlord or his agents diligently made their rounds at least weekly checking on the crops. Occasionally, the landlord or his agent lived on the place and kept a daily eye on the freedman's farming. If the tenant fell behind in his patch, the landlord might bring in additional laborers for a few days' work at the tenant's expense, force freedmen on the advanced patches to help their less productive neighbors, or even replace the tenants with new groups working under new contracts. Such stipulations appeared in both share and fixed rent tenancy contracts often enough to indicate that landlords, at least in the 1880s, saw little difference between the two systems insofar as these risks were concerned.

Nor is it correct to think that the choice between the two systems reflected the ease with which sharecropping facilitated aggregate risk reduction.[57] In the first place, there is little evidence that sharecropping rather than fixed rents or fixed wages better enabled the parties involved to renegotiate production plans during the year "in response to deviations from initial expectations" resulting from weather conditions and so on. With sharecropping, the landlord and tenant theoretically had a mutual incentive to alter previously agreed-upon production plans to take advantage of possible windfalls or to offset threats to production. In reality, however, the landlord's position and the nature of the situation hardly required shares as a mechanism of response. One is reminded here of the widow, Mrs. Jane E. B. Conner, who leased one of her Concordia plantations to several groups of freedmen for 175 bales of cotton in 1875. Although the freedmen were fixed rent tenants, the contract stipulations recorded in the mortgage records gave Mrs. Conner's agents all authority to insure that the rent crop be made, even to the point of putting off any "obnoxious freedmen."[58]

The case of J. A. Gillespie's contract with his fixed rent tenants on Indian Village is an especially pertinent example. Gillespie's freedmen tenants had refused to accept his demanded 50-bale rent for Indian Village in 1873. After much haggling, Gillespie accepted their offer of 36 bales, but only after considering the use of "chinamen" as tenants. The same terms were agreed to for the next year. Then the floods hit. Indian Village, inundated with water, produced a total crop of 15 bales. The tenants petitioned Gillespie to remit their lease agreement, which he did providing that they "repair and rebuild all of the buildings on the plantations."[59]

In the second place, the low levels of skill required in cotton production in the postbellum South makes it absurd to think of anyone taking advantage of windfall profits through alternations in existing contract provisions. Windfalls could hardly have occurred. Nothing related to new technology or information was realistically a possibility. Similarly, although flooding and other natural disasters were frequent enough, little could be done under most conditions to alter the situation. If anything could be accomplished, such as building a makeshift levee, all hands— tenants and croppers and wage hands alike—pitched in or else faced the wrath of the landlord or merchant supplier. None had any choice in the matter, since all equally depended on the store for their daily subsistence.

Studies of other southern areas have shown that the explanation for the contract choice reflected the differing costs of the "landlords' enforcement of labor services." The argument is that a share rent tenant would not labor as much as a fixed renter. Thus, although the planter had contractual ability to insure the crop regardless of the terms of rent, he might not have been willing or able always to pay the costs of supervision required of share tenants. He would have preferred the lower supervisory costs of a fixed rent system, especially on large plantations. Such theoretical explanations would suggest the dominance of fixed rent tenancy in the Natchez District, since the landlord's supervision costs would be less with greater independence for the freedmen. The only plausible explanation for the dominance of sharecropping according to the above reasoning is that the added costs of supervision for sharecropping, if there were any, were costs the planter was willing to bear. Such logic is nearly useless, however, since it tells us little beyond what was a possibility. It is insightful, however, if the planter's willingness to bear these added costs resulted from his refusal to believe that blacks were capable of working as self-directed tenants.[60]

It is possible, however, that the dominance of sharecropping reflected the supplier's belief that it was a better means of protecting his investments. On the face of it, there was no reason for the merchant to have preferred doing business with a share tenant over a fixed rent tenant, since by 1880 any advances made were protected with a crop lien that gave the merchant first claim on the harvest. Moreover, with no real distinction existing between the landlord's ability to supervise the crop's progress, both arrangements forced the freedmen to concentrate on growing cotton to secure the needed advances for farming and living. Research in the lien records indicates little hesitancy by suppliers to deal with fixed rent tenants on practically the same basis as with sharecroppers. Merchants David Singleton and David Young, for example, supplied fifty groups of freedmen on four plantations in 1874, with no apparent dismay over their tenure status. The two merchants saw little difference between the two types of tenants, viewing both fixed rent tenants and sharecroppers as "laborers" with whom they agreed "to furnish...as partners in planting with supplies to make a crop." The fact that they were alluded to as "partners in planting" meant only that rents due the landlord would be drawn from the total crops rather than from the supplier's share exclusively.[61]

But differences existed other than the form of rent between the two forms of tenancy. For one thing, the fixed renters were more diversified and probably not as poor as the share renters. A close analysis of the manuscript census for 1880 reveals that 85 percent of the district's fixed rent tenants grew corn in comparison to 50 percent of its sharecroppers, that 76 percent of its fixed rent tenants owned tools in comparison to 50 percent of the croppers, and that 82 percent of the renters owned animals in comparison to 67 percent of the croppers. These figures, of course, suggest some correlation between tenure status and assets. Even more revealing is the fact that, according to the mortgage records, the fixed rent tenants paid less rent than the croppers. In situations where the cropper furnished the tools and animals and feed, he usually paid one-third of the crop as rent. If the cropper furnished nothing but his labor he paid half the crop as rent. Since the average number of acres improved per sharecropping farm in 1880 was 18.6 acres, with an average production of 12.6 bales per farm, the typical half-cropper paid 6.3 bales of cotton, while the third-croppers paid 4.2 bales of cotton in rent. Or, to put it another way, the typical third-cropper in 1880 paid 0.22 bales of cotton per acre rented. In contrast, the typical fixed rent tenant listed in the mortgage records paid no more than 2.7 bales per farm or 0.12 bales of cotton per acre rented.[62]

The diversity and lower per acre rents associated with fixed rent tenancy indicate that the fixed renter offered advantages to planters which undoubtedly prompted a few to accept lower rents for their lands. That fixed rent tenants tended to own their own tools and animals and grew corn more often than not in comparison to the croppers is certainly a description of a relative independence that possibly worked to a planter's benefit. Simply stated, the advantage may have been the fixed rent tenant's ability to function independently of the merchant for supplies. Thus he was less likely to have his crop tied up with crop liens. The merchant's lien for supplies took privilege over the landlord's lien for rents and thus provided some incentive for landlords to accept fixed rents from relatively independent freedmen. These rents were perhaps guaranteed if the tenant could stay out of debt to the local store. In this way the freedman, if truly independent of the merchant, could diversify his farming after planting enough to make the rent. The fact that the average size of the fixed tenant's farm exceeded that of the sharecropper's patch supports this contention if the additional acres were in vegetable gardens and market crops other than cotton.[63] Although such independent tenants were few and scattered, they did exist as indicated by the practice of the prominent Surget family in renting several farms to freedmen in the 1870s for set cash rents, with half the amount payable in advance.[64]

Nothing above fully explains why sharecropping of whatever form dominated over fixed rent tenancy. Little real difference existed between the two except in the matter of capital assets in a relatively few cases. Neither merchants nor landlords had any clear reason for preferring one system over the other, except in those few cases where the tenant's assets assured the landlord a fixed rent, since both were able to exercise sufficient supervision in either tenure choice to protect their interests. Moreover, the freedmen's poverty or lack of assets may have at first influenced their choice towards shares because it better enabled them to reap the benefits of high cotton prices. By 1880, however, the nagging dependence of share tenants and fixed renters alike on the merchant undoubtedly obscured such market orientation. Even those tenants who owned tools and animals and paid a fixed rent, if the lien records can be believed, were indebted for them to the merchant rather than owning them clear.

Their situation was so frustrating by 1879 that hundreds of the district's freedmen—croppers and fixed rent tenants alike—made ready to journey to Kansas in what amounted to a strike at the economic trap in which they found themselves. Although observers described the unrest as the irrational "Kansas Fever," it was more accurately the freedmen's expressed recognition of how much their hopes for sharecropping had gone unfulfilled. The rumor that Kansas promised free land meant that none had anything to lose by the venture. It was only through the combined use of propaganda, force, subterfuge, and especially the promise of planters and supplymen throughout the district to cancel all debts that the disgruntled freedmen agreed to abandon the Kansas dream and return to work. Although the concessions only momentarily relieved the situation, since most hands quickly accrued new debts for supplies and seeds, the episode confirms the extent to which freedmen viewed share tenancy and fixed tenancy as much the same system. In each, they generally functioned as indebted laborers rather than entrepreneurs and as dissimilar farm hands rather than differentiated farmers at distinct steps on an agricultural ladder leading to farm ownership.[65]

It seems then that sharecropping dominated over fixed rent tenancy because fixed rent contracts offered no particular advantages over cropping. In this sense, the tenure choice reflected mainly the historical circumstance that sharecropping had evolved directly from share wages because of the freedman's vigorous rejection of low wages and gang working conditions. In the cases where fixed renting did prevail, the choice stemmed partly from the freedman's competitiveness and capital assets but chiefly from individual decisions for which no clear patterns emerge. Some landlords simply preferred fixed renting for personal reasons. The reality of this interpretation seems clear once it is recognized that the eventual transition to fixed renting from sharecropping in the 1880s and 1890s coincided with the bottoming out of cotton prices. Once landlords and suppliers found that a share of the crop brought less income each succeeding year due to the price declines of cotton, they began dealing with the freedmen on a fixed rent basis usually payable in cotton. But this eventuality had little meaning to landlords in the Natchez District in the immediate postwar era.[66]

CONCLUSION

In conclusion, although scholars have correctly suspected that all of the participants played significant roles in the system's development, it seems that sharecropping originated largely in the freedman's insistence upon the arrangement. The freedman's determination generally reflected his understanding that shares enabled him to counter the low wage scales set early in the postwar period and facilitated his attempts to function as independently of the planter as possible. This latter aspect of the freedman's determination complemented his resolve to undermine the gang labor conditions of slavery, although his hatred of closely supervised labor was not directly responsible for the origins of

Plate 8. Leaving for Kansas. From *An Album of Reconstruction* by William Loren Katz, copyright 1974. Used by permission of Franklin Watt, Inc.

sharecropping. The two were separate aspects of the freedman's new-found liberty of contract resulting from the intensive competition for his labor.

This is not to say that planters and merchants played less than crucial roles. Indeed, the move to shares occurred only because the system enabled both groups to live with and profit from the system. Planters and merchants found sharecropping acceptable mainly because of the lack of alternatives but partly because it offered few impediments to doing a prosperous supply business, on the one hand, and because it especially enabled landlords, on the other hand, to lower their production costs by bringing the merchant into the operation as a partner of sorts. In neither case did the planters or merchants initiate sharecropping in hopes (1) of increasing the freedman's personal efficiency, (2) of sharing with labor the risks of production, or (3) of reducing the costs of supervising wage labor. Numerous planters undoubtedly believed that shares were essential if gang labor could not be had, but the majority of planters who accepted shares did so to obtain hands on terms set largely by the freedmen. This suggests that only a few planters understood that the technology of cotton production seldom demanded gang labor conditions except as a mechanism of social control.

Many of the reasons given above for the dominance of sharecropping over fixed renting contradict the usual interpretations about the determinants of tenure choice. For most scholars of the subject (historians and economists alike), the choice between half shares, quarter shares, third shares and cash or value tenancy depended on the amount of supervision afforded by the landlord or merchant. According to this reasoning, fixed rent tenants, bearing all of the risks of farming, would not accept the kind of supervision exercised over hired hands or sharecroppers, who bore little or no risks, since they contributed neither tools nor lands. Planters and landlords preferred sharecropping because they could better manage, and freedmen had little say in the matter unless they contributed a portion of the tools or animals.[67]

Such motivations, however, were seldom responsible for the origins and dominance of sharecropping in the Natchez District. Risks were similar among the tenure choices because few tenants were free of debts for the tools and animals they contributed to the production process. The key issue was not the degree of managerial supervision because neither the landlord nor the merchant supervised the actual cultivation of cotton in either share or fixed tenancy. Yet, in all cases landlords and merchants exercised nearly total control over the crucial decisions of crop production. If anything, freedmen who farmed on a fixed rent basis did so because they could afford to and once they did, they tried to become as self-sufficient as possible. The real difference between sharecropping and fixed rent tenancy was the greater degree of self-sufficiency possible in the latter case. This had nothing to do with the issue of supervision, except that freedmen with corn might to some degree avoid the control of their food supply. Sharecropping predominated over fixed renting because some freedmen could afford to rent. Others rented because some planters preferred a fixed rent for reasons of their own that may be too numerous and vague even to estimate. But it should be understood that in most cases freedmen were free of immediate and daily supervision because they had insisted upon this condition as the basic term of their employment. On the other hand, few freedmen exercised much independence in the production decisions of what to plant, when to plant, where to plant, or how to plant, unless their assets freed them of dependency upon the store.

Much of the confusion about the role of risk, supervision, and the so-called incentives for labor in the determination of the tenure choice flows from a misreading of the nature of southern farming under sharecropping. Historical economist Joseph D. Reid has suggested that less experienced farmers (freedmen) always preferred sharecropping because of the assistance they might receive from their more knowledgeable landlords.[68] Yet, the experience of Nate Shaw in Alabama in the years before World War I suggests that Reid's speculations are far removed from the reality of southern agriculture. Shaw hated sharecropping because he was not free to farm as he wanted to farm. It was not that his landlord supervised his daily work, but that as a sharecropper with neither mules nor capital he farmed the poorest lands with scant assistance from the landlord or his merchant supplier. Fixed renting, on the other hand, meant that Shaw could work his mule as long and as hard as needed because it was his mule. He could haul his cotton to the gin of his choice, and he could hitch the animal up for a Sunday drive without asking permission from its owner. This was the real difference between sharecropping and renting. Supervision, risks, incentives, and information costs were all part of the picture but not in the ways suggested by the theoretical view of the matter.[69]

In the Natchez District, sharecropping predominated as the main tenure mode until the twentieth century when it was replaced by cash or fixed rent tenancy. Davis, Gardner, and Gardner have found that nearly 57 percent of the tenants in Adams County were cash tenants in the 1930s. Why this system emerged is unclear. The above authors suggest that landlords were unwilling to supervise the tenants and much preferred a cash rent arrangement. Yet, there is substantial evidence that "even most of those landlords who lived on their farms maintained little or no supervision of the actual work of their sharecroppers."[70] It is more likely that the system shifted from share renting to cash tenancy as a means of better insuring the landlord a maximum income. In times of declining cotton prices, sharecropping meant less income for the landlord as the value of their share declined. In few cases did the landlords supervise the actual cultivation of cotton, and the reasons for this may be found in the circumstances surrounding the system's origins in the 1870s. Some landlords were unwilling to supervise their labor, many were absentee-owners, but, more importantly, freedmen had moved to sharecropping to avoid such supervision. It was as simple as that.

NOTES

1. U.S. Census (1880), Manuscript Population and Agricultural Schedules, Adams County, Mississipi, and Concordia Parish, Louisiana. There were essentially three kinds of sharecroppers: those paid wages each month and a small share of the plantation profits, or a share of the crop (usually one-tenth) at year's end; those workers supplying only their labor and paid between one-third to one-half of the crop made; and those supplying their labor as well as equipment or tools or animals to farm and retaining one-half to two-thirds of the crop made. The third group was often thought of as share tenants paying one-third of the crops as rent.

2. Registry of Leased Plantations, Records of the Bureau of Refugees, Freedmen, and Abandoned Lands (hereinafter cited as BRFAL), Record Group 105: 79-81.

3. See J. A. Gillespie Papers, Louisiana State University, Baton Rouge, Louisiana, for a good example of the move from slavery to wage labor, from wage labor to sharecropping, and from sharecropping to fixed rent tenancy.

4. See William Newton Mercer Papers, Louisiana State University, Baton Rouge, Louisiana.

5. William B. Shields to William Newton Mercer, December 4, 1866, Mercer papers.

6. Ibid., November 28, 1866.

7. Ibid., February 13, 1867.

8. Ibid., Mercer did manage, however, to convince some of his laborers to work for a smaller monthly wage plus a share of the crops.

9. Ibid., November 28, 1866.

10. Ibid., March 27, 1867, May 29, 1867.

11. Ibid., January 6, 1867.

12. Ibid., October 27, 1867.

13. Ibid., December 1, 1866.

14. Ibid., January 6, 1867.

15. Ibid., October 27, 1869, November 8, 1869.

16. See Lemuel Parker Conner Family Papers, Louisiana State University, Baton Rouge, Louisiana.

17. Ibid., see Undated Testimony.

18. Ibid., Elizabeth Francis Conner to L. P. Conner, July 11, 1863.

19. Ibid., Contract, March 13, 1865.

20. Ibid., F. Conner to L. P. Conner, August 21, 1866.

21. Ibid., L. P. Conner to Elizabeth Francis Conner, September 29, 1867.

22. Ibid., A. E. Conner to L. P. Conner, November 7, 1874.

23. Ibid., February 18, 1880.

24. Shields to Mercer, November 28, 1866, December 12, 1866, November 17, 1869, Mercer Papers. The freedmen working for Mercer preferred shares over fixed wages in the early years of Reconstruction. They were so market oriented (if that is the correct phrase to use) that Shields was forced to bring evidence of the selling price of cotton to the hands at settlement time. Shields later commented on their concern with the price of cotton in 1869 and suggested to Mercer that a fall in prices "would cause many freedmen to change their plans."

25. Ibid., December 4, 1866.

26. Ibid., October 13, 1869.

27. Ibid., July 10, 1866.

28. W. Lowenberg, Monthly Report, April 1865, BRFAL, Record Group 105.

29. W. F. Wheeler, Monthly Report, March 1865, BRFAL, Record Group 105.

30. J. F. Evans, Monthly Report, May 1865, BRFAL, Record Group 105.

31. Monthly Reports, January-December 1865, BRFAL, Record Group 105.

32. The bound mortgage records housed in the courthouses at Vidalia, Louisiana, and Natchez, Mississippi, are the best sources for detailing the varied contractual arrangements between freedmen and planters in the Natchez District in 1865, 1866, and 1867.

33. Shields to Mercer, January 1, 1867, Mercer Papers.

34. Ibid., May 22, 1867.

35. *Natchez Democrat*, December 23, 1865, p. 2; U.S. Congress, House, Testimony of Dr. James M. Turner, March 14, 1866, Joint Committee on Reconstruction, 39th Cong., 1st Sess., 1866, *H. Report* 30, Pt. 4, pp. 127-28.

36. See James M. McPherson, *The Negro's Civil War* (Vintage Ed., New York: Random House, 1965), pp. 142-244; Bell Irvin Wiley, *Southern Negroes, 1861-1865* (New Haven, Conn.: Yale University Press, 1938), pp. 295-344.

37. See John Eaton, *Grant, Lincoln, and the Freedmen* (New York: Longmans, Green & Co., 1907), pp. 216-17; E. Stanton to O. O. Howard, October 24, 1863, BRFAL, Record Group 752.

38. G. D. Reynolds to W. E. Strong, March 25, 1865, BRFAL, Record Group 105.

39. J. A. Gillespie to W.A.K. Farrar, April 26, 1877, Gillespie Papers.

40. Farrar Conner to Lemuel Conner, September 27, 1866, Conner Papers.

41. Shields to Mercer, November 8, 1869, Mercer Papers.

42. Postlethwaite to Gillespie, June 27, 1867, Gillespie Papers.

43. This is in contrast to the conventional notion that planters actually tried a viable wage system after the war. In fact, the wages offered, along with the working conditions employed, were never adequate. Thus, it is wrongheaded to think that the system of wages failed to attract laborers. A realistic wage system was never tried. For a succinct statement of the conventional view, see Jay R. Mandle, *The Roots of Black Poverty: The Southern Plantation Economy After the Civil War* (Durham, N.C.: Duke University Press, 1978): 17-21.

44. Benjamin R. Teele, "Natchez Planter Gives Up," *The Cultivator and Country Gentlemen* 31 (February 1868), p. 128.

45. F. Conner to L. P. Conner, August 21, 1866, Conner Papers.

46. See Stephen Duncan, Jr., to Stephen Duncan, Sr., January 11, 1863, Duncan Family Papers, Louisiana State University, Baton Rouge, Louisiana.

47. See Gillespie Papers.

48. Ibid., Contracts, January 1, 1868, May 5, 1868.

49. Ibid., Contracts, January 1869.

50. Ibid., Contracts, January 20, 1870, September 28, 1870, January 1, 1871, January 12, 1871, January 10, 1973, January 4, 1874, February 24, 1875, January 6, 1885, January 8, 1885, January 3, 1886, January 9, 1890. See also the Liens and Mortgage Records, 1865-1900, for Adams County, Mississippi, and Concordia Parish, Louisiana.

51. Contracts between David Williams and Freedmen on the Overton and Siego Tracts, February 1, 1870, Liens and Mortgage Records, Adams County, Mississippi, Office of Records, Natchez, Mississippi.

52. Ibid., Contracts between S. G. Kennedy and Freedmen, March 25, 1870.

53. Contracts between A. Smart and Robert Bradley and Joseph Baldwin (f), March 27, 1880, Liens and Mortgage Records, Concordia parish, Louisiana, Parish Court House, Vidalia, Louisiana.

54. See "Rules for Working Freedmen," February 22, 1864, W. J. Minor Papers, Louisiana State University, Baton Rouge, Louisiana. Compare these rules with his "Rules for Working Waterloo Plantation" in 1861, as found in his diary:

Overseer must (1) treat Negroes with kindness and humanity—when sick they must have every attention and see that Doc's directions be adhered to. (2) See that all hands are at work as soon as they can be—give particular attention to hands in field. (3) Must not strike Negroes with anything but his whip, except in self defense. Must not cut the skin when punishing—not use abusive language as it makes them unhappy and sometimes to run. (4) Examine quarters after ringing of the bell to see if Negroes are all at home at night. Require drivers to report absentees every morning. (5) Negroes should retire to cabins by 9:15. (6) Negroes cannot leave without permit. (7) Negroes are not allowed to beat wives. (8) Divorce and remarriage require 25 lashes—6 months notice for divorce. (9) Keep record of quantity and condition of stock. (10) Not allow Negroes to swear, do anything disrespectful, make noise in quarters, nor talk loudly while at work, nor allow quarreling or fighting. (11) Do not allow Negroes to keep or use spiritous liquors. (12) See that rations are properly delivered. (13) Visit stables every day at 12:00 and at night. Note—Negroes are in the habit of regulating the depth of the plow by the back hand, thereby throwing the whole weight of the draft on the back of the animal working. They are also fond of running up on the tracks of the animals in such a manner as to prevent them from throwing their weight into the collar. (14) See that houses and quarters are cleaned up once a week—especially the back yards. (15) See that Negroes dress clean every Sunday. (16) See that mechanics not make or sell any of their own work without permission. (17) Do not allow mechanics to strike or mistreat hands under them. (18) When necessary to punish, inflict it in a serious, firm and gentlemanly manner. (19) Farming utensils are regularly put away and in order. (20) See that all ditches drain well as the Negroes will not work them as he works the crops so (keep them clear all year). (21) He must not allow the Negroes to use the horses, carts, or wagons without specific permission—neither must he allow the hands to ride to and from the fields in carts. When they may be going or coming at the same time serious accidents have occurred from this habit. (22) Record all births and deaths. (23) Record all receipts and shipments. (24) Keep plantation books with notes of interest. (25) See that seamstress makes the clothes strong and neat. (26) Preserve all manure. (27) Never leave the plantation.

55. Contracts between Martha Claiborne and C. Carroll, January 1, 1870, Liens and Mortgage Records, Adams County, Mississippi.

56. Ibid., Contracts between Thomas B. Shields and Freedmen, April 2, 1870.

57. See Robert Higgs, "Patterns of Farm Rental in the Georgia Cotton Belt, 1880-1900," *Journal of Economic History* 34 (June 1974): 468-82; Robert Higgs, "Race, Tenure, and Resource Allocation in Southern Agriculture, 1865-1910," *Journal of Economic History* 33 (March 1973: 149-69; Joseph D. Reid, Jr., "Sharecropping as an Understandable Market Response: The Post-Bellum South," *Journal of Economic History* 33 (March 1973), pp. 106-30; Roger L. Ransom and Richard Sutch, "The Ex-Slave in the Post-Bellum South: A Study of the Economic Impact of Racism in a Market Environment," *Journal of Economic History* 33 (March 1973), pp. 131-48.

58. Contracts, February 18, 1875, Liens and Mortgage Records, Concordia Parish, Louisiana, Parish Court House, Vidalia, Louisiana.

59. See William Jett to J. A. Gillespie, November 10, 1872, November 17, 1872, November 16, 1873, April 1, 1874. A. L. Farrar to J. A. Gillespie, February 28, 1873, Gillespie Papers.

60. See Higgs, "Patterns of Farm Rental in the Georgia Cotton Belt." Interestingly enough, Higgs passes over the possibility that the ownership of plantations by merchants or absentee owners may be of some importance because such data may not be incorporated into his statistical analysis.

61. Contract, January 24, 1874, Liens and Mortgage Records, Concordia Parish, Louisiana, Parish Court House, Vidalia, Louisiana.

62. U.S. Census (1880), Manuscript Population and Agricultural Schedules, Adams County, Mississippi, Concordia Parish, Louisiana.

63. Ibid. The additional acreage might also have reflected the quality of the soil rented relative to that farmed by sharecroppers. Inferior soils possibly required a greater number of acres to make the rent.

64. Contracts, January 1870, Liens and Mortgage Records, Adams County, Mississippi, Office of Records, Natchez, Mississippi.

65. William Ivy Hair, *Bourbonism and Agrarian Protest: Louisiana Politics, 1877-1900* (Baton Rouge, La.: Louisiana State University Press, 1969), pp. 83-106; Nell I. Painter, *Exodusters: Black Migration to Kansas after Reconstruction* (New York: W. W. Norton Co, 1976), pp. 196-201.

66. See Allison Davis, Burleigh B. Gardner, and Mary R. Gardner, *Deep South: A Social Anthropological Study of Caste and Class* (Chicago: University of Chicago Press, 1941), p. 288.

67. For the most recent contribution to this debate, see William Parker, "The South in the National Economy, 1865-1970," *Southern Economic Journal* 46 (April 1980), pp. 1019-48.

68. Joseph D. Reid, Jr., "The Theory of Share Tenancy Revisited— Again," *Journal of Political Economy* 85 (April 1977), p. 406.

69. Theodore Rosengarten, *All God's Dangers: The Life of Nate Shaw* (New York: Avon Books, 1975), pp. 104-361.

70. Davis, Gardner, and Gardner, *Deep South*, pp. 255-342.

• 5 •

PLANTERS AND MERCHANTS

Robert Somers tells of meeting a Jewish merchant in southern Mississippi in 1870 who supplied sharecroppers and planters at "100 percent of retail profit." "Mr. Solomon" protected his advances through a lien on the crops and, in the case of his freedmen debtors, regularly sent riders out to check on the condition of his security. This practice had partially usurped the managerial role of the antebellum planter by the end of that decade. But in Solomon's opinion, such high profits existed mainly in "de books" and not in "de pocket" since "de white planter is very poor, and de negro, who sometimes raises crops for himself, is very idle, and knows no accounts." The idea of "white christian people, possessors of large landed estates," in "bondage" to a Jew greatly amused Somers, and we are left with the picture of a good-natured, heavily accented, ingenious Jewish trader lording over a domain of penniless whites and idle freedmen.[1] Solomon might have been one of several score furnishing merchants in the Natchez area, possibly a Mr. Solomon Asher of Asher and Wolf & Company. It is important to understand his function and how it affected the area's economy.

The role merchants came to play resulted primarily, of course, from the abrupt entry of a great number of formerly enslaved people into the demand end of the retail market. But the ensuing relationship of freedmen to merchants as customers and debtors was not immediate. Few emancipated slaves were able to direct much of their earnings to local merchants during or shortly after the war years. They worked on leased government plantations at subsistence rations, they could not leave without the planter's permission, and they were paid low wages in partial allotments. Therefore, most freedmen in 1864 were forced to consume their earnings before payday in the plantation store. The procedure was simple enough. Some planters paid their hands at day's end with tickets that were redeemable on payday for greenbacks or could be spent in the plantation store at any time. Needless to say, the tickets were seldom negotiable beyond the plantation. Most planters, however, kept account of a hand's daily purchases and earnings to make sure that purchases did not exceed earnings by any substantial margin. This practice continued after the war with little change except that by then some planters felt it was to their advantage to let their hands end the year a little in debt to the store.[2]

But even under such conditions, many freedmen did some marketing beyond the plantation store. Peddlers and hucksters, for instance, plied the safer roads and waters in Adams County in 1864 exchanging bonnets, flour, and notions with the freedmen in return for chickens, hogs, fish, and even cotton. Few questions were asked about the customer's legality of ownership.[3] It was no exaggeration for Natchez planters to complain of losing everything to freedmen and soldiers that was not nailed down.[4] The unsettled times also enabled freedmen to slip away from the plantation after dark for a little nocturnal business with petty entrepreneurs in the area.[5] Still others, leaving the plantation completely, took to the woods where a day's labor with ax and saw or a few traps secured them enough cash for alcohol and tobacco.[6] Some undoubtedly managed to hide away enough wage tickets to obtain a few greenbacks for a day's spending in Natchez. This fact was of no little chagrin to one old-timer who claimed that neighborhood blacks had more dollars in their pockets than any white person in town.[7]

Although planters continued to hold the lion's share of the freedman's business, several developments were underway that were to affect seriously the planter's role as chief source of supply. The least of these involved the Freedmen's Bureau's decision to enforce a claim on the planter's crops in favor of the freedman's wage bill. This crop lien finally became law in the area in 1867.[8] A laborer's lien on the crop superseded a landlord's hold for rent or a supplier's hold for advances from sources other than the planter. Most merchants understood, however, that the freedman's lien hardly insured the merchant of payment for supplies advanced, since the planter controlled the crop's sale, and no guarantee existed that laborers could prove they had wages coming at the end of the year. Accordingly, few merchants advanced supplies to freedmen unless the planter himself endorsed the transaction, and few planters able to supply freedmen saw much advantage to signing these contracts.

More importantly, many planters were unable to survive financially, let alone to continue supplying freedmen on their own accounts through the plantation store after 1867. The case of J. D. Shields was typical.[9] Shields, confident of his future in 1861, had purchased his partner's half of the Pecano plantation in Concordia for $84,000, with a downpayment of $14,000, secured from his New Orleans factors, with the remainder due in seven annual payments of $10,000 at 8 percent interest. Shield's former partner, a Mr. Kebbe, then bought a plantation in Arkansas from J. R. Jones for $84,000 giving Shields's notes as collateral to secure a mortgage on the Arkansas place. Unable, during the war, to pay either the annual installments or the interest, Shields found himself in arrears for over $40,000 in 1865.[10] The dissolution of Shields's old factorage house in New Orleans left the planter with no alternative but to plead with his creditors for more time.[11] Explaining that he had failed to raise the money from either local sources or from friends in New York, Shields suggested either leasing the place and giving a two-thirds share of the rent to Jones and Kebbe or working the plantation himself with all profits going to settle his accounts. Jones accepted the latter suggestion with the stipulation that he be allowed to buy half the plantation at $15 an acre with the proceeds going to cover the old planter's back debts to Kebbe. Both then agreed to work the place, sharing costs and profits.[12]

Hoping thus to start anew, Shields approached the reorganized factorage house of Buckner and Company for supplies, offering a mortgage on his half of the crops to secure any advances made. Successful in securing from his old factors enough supplies to start planting, Shields hired a manager for $2,500, stocked his plantation store, employed many of his former slaves for fixed wages and rations, and once again set to planting. Everything depended on that year's venture.[13]

But 1867 was not a good year for the new partners. In the first place, as indicated in a note from Shields's manager, things got off to a bad start with the freedmen:

The hands still refuse to contract and say they won't until you and Jones write the contract. I have done my best but the fools won't listen to anything I have to say. I think they want to force you to let them raise cotton. I do wish Jones would come. I feel he won't bring half hands enough.[14]

Although it is impossible to tell from the correspondence how this affair ended, it is clear that the freedmen were less corrigible that year than ever before. In addition, the spring floods broke through the dikes damaged during the war and washed out most of the newly planted cotton. By the time the waters had receded enough for a second planting, there was little hope that much could be made. As a result, New Orleans merchants and factors refused to advance to planters in the area, causing many to lay off their hands and give up planting in despair. Reduced to feeding swamp moss and bushes to his stock, Shields managed to hold on through the summer only to face complete ruin in August as the so-called army worm descended hungrily on his crop. Shortly thereafter, the old planter gave orders to kill the stock for food and give any remaining cotton to the freedmen for their wages.[15] His valiant try had ended in failure.

Shields's experience was a typical one in the area, though the damage done by flooding and the army worm in 1866 and 1867 would have been easier to bear if the district's lands had been free of debts and mortgages. A great many planters had given mortgages on their plantations in order to secure enough supplies to begin planting in 1865. Most notes were given to New Orleans factors or eastern commission merchants, and the record books still housed in the district's courthouses illustrate the extent to which the war ended the antebellum practice among factors of advancing credit secured only by the growing crop or the planter's word. Sarah Welch, for example, mortgaged her Grand Cut-Off plantation in Concordia to the New Orleans house of Buckner and Newman for supplies advanced and to be advanced amounting to $24,000. Welch had piled up over $16,000 in debts to the New Orleans firm prior to the war, and her factors refused to advance more without a mortgage.[16] Even the prestigious John Minor, one of the wealthiest of the antebellum planters, was forced by the hard times to mortgage his Palo Alto plantation to Leverich and Company, commission merchants in New York who held almost $15,000 of the planter's promissory notes, for an additional loan of money at 7 percent per annum.[17] In one sense, the giving of a mortgage to coastal factors symbolized the passing of an age.[18]

The placing of liens on lands that had never held a mortgage became such common practice that few families were spared. William Gillespie mortgaged Frogmore to Buckner and Newman; George W. Surget his Coosa plantation; and Wesley Conner did the same to Roseland, as did Elias J. Hoover his Union Point.[19] But by that year's end, 1866, it was increasingly difficult to find continued support from coastal lenders. Unable to pay their interest and debts, Natchez District planters turned to one another for help. William Conner secured a backer in the widow of old-time planter A. B. Baker, in return for notes on his Hayhogan plantation.[20] The Percys mortgaged their Zeanov place, 8 miles above Natchez, to William B. Duncan for $5,000, which they desperately needed to begin planting.[21] Henry Chotard turned to his extended family for a loan of $22,000 for which he signed over Sumerset plantation to his obliging sisters and aunts.[22] Russell Conner, unable to settle the preceding year's supply debts to his New Orleans factors, found his neighbor, James A. Gillespie, willing to endorse his extension in return for a trust deed on Buckely.[23] Most illustrative of this development was the plight of Henry S. Metcalf, one of the district's wealthiest antebellum "Nabobs" and the man one neighbor's overseer had accused of contracting with blacks on any terms demanded. In April, Metcalf borrowed $20,000 from his brother, John, of New York, to plant his Grove plantation, giving a lien on all tools, animals, and crops thereon as security. Three months later, the old planter, seemingly desperate for funds, borrowed a total of $26,347 in a complex package of twenty-four promissory notes payable over the next six years. His creditors included neighbors like the still wealthy Davis family, a Mr. Charles Weeks, a New Orleans mercantile house, and several close family friends. In return, Metcalf "granted to the above creditors in mortgagee" his Gem, Woodland, Montrose, York, and Bourbon plantations, totaling 7,233 acres.[24] The old planter needed the funds to settle $48,305 in debts which he and his brother had incurred during the war with the New Orleans firm of Buckner and Newman. Metcalf later contended that these debts were in inflated Confederate money advanced for supplies for making a crop upon which Buckner and Newman were "to get the business" and were computed at a higher rate of interest than the law allowed when not stipulated in writing. Instead of settling his accounts, Metcalf apparently used the new funds to try planting once again in hopes of making enough to finally clear himself. But it was hopeless. In

1869, having met neither his old creditor's nor his family's notes, Metcalf sued for bankruptcy and, along with his family, contested the heirs of his New Orleans factors for title to the plantations claimed by both sets of creditors. The courts ruled in favor of the New Orleans heir, Henry S. Buckner, who bought the four plantations for $18,000.[25]

Not surprisingly, planters in similar predicaments, those heavily mortgaged and in debt, had little alternative after 1867 but to sell or rent their plantations to outside investors. But the very conditions that drove planters to put their lands on the market had dried up the supply of takers among both outside sources of capital and local investors. The case of F. Lloyd King is illustrative.[26] A former Georgia planter ruined by the war, young King was hired by a New York firm to manage several plantations it had leased in the Natchez District. His employers wanted someone with planting experience to grow cotton on a large scale and yet innovative enough to meet the challenge of working imported German immigrants in the Louisiana swamps. Although King proved a competent and even venturesome manager, the experiment failed after two years. The problem had not stemmed from a lack of funds but from developments beyond the power of any one person to handle. It turned out that the German laborers were interested primarily in securing passage to America, and they promptly deserted at the first opportunity.[27] Yet, the supply of freedmen willing to work in the swamps for wages and rations was never enough to staff all the plantations leased. In addition, King had a difficult time with his overseers and managers. The worst of the lot were drifters, and even the best of them paid little attention to their jobs or they treated the freedmen as slaves rather than as hired laborers. On a visit to the Tanglewood plantation, King found his manager off on a visit, the freedmen ready to revolt, and only two days' rations left on the place.[28] Finally, the spring floods of 1866 and 1867 destroyed the corn and cotton crops in spite of his almost superhuman efforts to save something from the water and the ensuing worms. Although King did make a life for himself in the Natchez District, but not in cotton planting, his employers found the sequence of events too much to bear and withdrew all investments from the area after 1867.

Nor did the prospects of attracting buyers or renters appear much brighter for planters in the dry uplands of Adams County after 1867. Too many of the country's plantations suffered from such advanced states of soil exhaustion as to warrant but little serious consideration for investment by speculators hoping to make their fortune on a bumper crop. In fact, few Adams County planters experienced that brief period of euphoria felt by those bottom land Concordia planters who had cashed in on the mania for swamp lands among outside speculators and capitalists.

Confronted with a smaller demand for plantations by men of capital, a difficult labor situation, and a tight credit market, many Natchez District planters were unable or unwilling after 1867 to assume planting on their own accounts. As a result, planters began leasing their lands on sacrifice terms to white tenants of little or moderate means. A. B. Kembly, for instance, rented a portion of James Surget's plantation, Waterloo, in Concordia for a share of the crop, employed freedmen on shares, and secured $800 in supplies from the firm of Meyer, Deutsche and Company of New Orleans.[29] Other lands were available throughout the district on similar terms. The *Natchez Democrat* listed at the end of 1867 over thirty plantations for rent in Concordia for a share of the crop, and one Natchez merchant, Morris Wexler, was selling lands on commission for only half the common harvest. In the latter case, Wexler also supplied the buyer in return for a mortgage on the remaining crops and land. Thus, a penniless individual could end up either owning a plantation after perhaps a single successful harvest or in debt to his supplier. Here the supplier probably gained the lands, possibly to his regret, through foreclosure and thus most likely promoted the arrangement with a new buyer the following year.[30] By these arrangements, Wexler undoubtedly hoped to find a buyer able to make a crop or a buyer solvent enough to withstand one or two poor ones.

But the seeming appeal of even these liberal arrangements was limited by the experience of the past few years. E. Jeffords leased the Nevitt plantation in 1866, with Meyer, Eiseman and Company of New Orleans to advance supplies and cash for making a crop through their agent, Louis Hart, a local Natchez merchant. The New Orleans house covered Jeffords's expenses each month, including lease installments, cash for wages, supplies, and whatever else Hart endorsed. Jeffords and Meyer agreed to share equally the gains and losses at year's end. Expenses in December amounted to $28,011.05, with total receipts of $9,380.59 for the sale of 45 bales of cotton, animals, tools, seed, and so on. Some additional income reduced the total loss for the year to

$18,562.22, of which both parties shared $9,280.11 each. Besides his labor, Jeffords had contributed $1,800 in cash, twenty-two mules, some tools, and a few supplies. When these items were deducted from his supply bill due Meyer, he ended the year owing the New Orleans firm $2,645.70. With this amount added to Meyer's share of the loss, the company was behind $11,925.80 for its partnership on the Nevitt plantation. The sources fail to explain what exactly had gone wrong, but such enterprise could ill afford to be repeated. Clearly, it was a disastrous experience for all the parties involved, except for the local merchant, through whom the advances had been channeled, and the landowner, assuming his rent was paid.[31]

With planters unable to carry such losses, fewer instances of this kind of planting partnership occurred. More commonly, planters began dealing directly with local merchants in ways unprecedented in antebellum times. The transition was subtle and is not to be thought of as the abrupt displacement of one group by another. Indeed, coastal supply houses continued to play an important role in the area's economy well into the twentieth century. But in their eagerness to secure business of every sort, local merchants proved worthy and successful competitors. From the moment the U.S. Army had occupied the district in 1863, merchants had swarmed into the area. Some were even then representatives of New Orleans firms attempting to obtain as much of the abandoned and picked cotton known to be ready for consignment.[32] Along with those who wanted only a permit to take the abandoned cotton to market came dozens of storemen, peddlers, hucksters, and merchants eager to supply the leased plantations with tools, seed, and rations.[33] The U.S. Army found the supply situation chaotic and moved forcefully to establish order.[34] Treasury agents issued permits on a random basis at first and then with an eye to favoring U.S. soldiers and those "old families" who in their poverty were reduced to the supply business.[35] In June 1864, the Treasury Department issued trade permits to seventy individuals in amounts ranging from $500 to $3,000 per month. Each was required to post bond.[36] By October, with the petitions for authorities to trade growing in number daily, the department limited the permits to the category of disabled U.S. soldiers. But this hardly stopped other shrewd aspirants. Jacob Arnett, for instance, a disabled soldier from Cincinnati, obtained a permit in April 1865 and then immediately joined in partnership with a local merchant, Max Lechner, who furnished the capital for their $3,000 a month supply business. Other soldiers did the same or sold their permits to entrepreneurs who then functioned with the power of attorney as supply merchants.[37] By the war's end, it is no exaggeration to say that the Natchez District was literally overrun with storemen.

For planters in the Natchez District, this horde of merchants was viewed as a mixed blessing. At first, planters preferred to obtain their supplies as in times past from their factors or local and well-established merchant dealers in Natchez. But the events described above soon found the once proud planters doing business with any one willing to accommodate their needs. Julia A. Nutt hesitated not at all to give a crop lien to local merchants, Ober and Atwater and Company, in return for $4,500 advanced payment with 8-percent interest.[38] She had little choice, alone as she was with four children to raise, her once wealthy husband dead, and nothing but the basement of the district's most imaginative plantation mansion, the unfinished Longwood, to call home.[39] Still others secured supplies from local merchants by selling them an interest in the coming crops at a price to be determined by the future market. For those dealing with local merchants in the above way, the supplies included everything from cash advances for wage payments and taxes, to seed, tools, and rations or food supplies. Those who could borrowed money and purchased supplies from competing merchants on a cash basis, sometimes borrowing the money from one source and buying from another.

By 1867, a new development began to take shape in the district. More and more planters were leasing lands to local merchants outright or to freedmen tenants, who in turn obtained the means of planting from neighborhood stores. In both cases, although the terms varied from share to fixed rent arrangements, in the crops planted, the rules of discipline, and the landlord's right of intervention, the merchant's role as the chief source of supply loomed large. Thus was established a necessary condition for the eventual decline of the landowner's function as an active on-the-place plantation manager or resident planter. Adams County planter Thomas R. Shields, for instance, after several years of working with wage laborers, leased his Hermitage plantation, including all implements and stock, for five years to several families of freedmen. The deal specified a rent of 18 bales of cotton, quarters for the freedmen in the plantation house, and removal of the landlord from the premises,

Plate 9. Longwood. Used by permission of Dr. Merle C. Nutt, P. E., author of "The Nutt Family Through the Years."

leaving only a mutually agreed-upon headman (with little supervisory authority) as a go-between. Shields not only abandoned his plantation residence but also gave up the supply business of his plantation store. Freedmen might still secure supplies and money from Shields at 10-percent interest or have him sell their crops for 2½-percent commission, but he fully expected them to obtain credits from local merchants and was resigned to collecting his interest and commission from "the merchants to whom the tenant's cotton might be sold or consigned."[40]

More typical yet was Martha Claiborne who rented over 200 acres of her Dumbarton plantation to six families for a one-third share of the crops.[41] Unlike Shields, Claiborne remained on the place and employed four other freedmen as laborers. She paid them half of the crops as wages with the stipulation that they furnish their own rations and share equally with her the expense of feeding the stock. Although Claiborne advanced supplies and money to several of her tenants, it may have been done under duress. The records of her neighbor, Henry K. Farror, indicate that some planters were unwilling to supply their tenants unless forced to do so in order to save the crop. Farror furnished his tenants with enough supplies to get a crop started, but once planting was underway he made advances only if they failed to obtain credit from other sources.[42]

Even J. A. Gillespie began to rely on local merchants in this way. As early as 1868, Gillespie's contracts show him moving away from the ration system, although he never completely abandoned his role as an adjunct supplier. Contracting with "Pepe Reilly and the undersigned," Gillespie's nephew indicated that bringing the merchant into the picture had become a widespread practice. In his case, the nephew refused to supply his freedmen tenants until midsummer and abruptly ended all advances precisely on December 18.[43]

Also common was the move to lease directly to local merchants. But it would be misleading to think of these merchants as active farmers or plantation managers, as might be assumed by their status as so-called planters. Instead, these arrangements enabled merchants to sublease to freedmen or to unite, in theory, in partnership with blacks who labored for a share of the crop and, in both cases, drew upon the merchant's store for supplies. In 1869, William F. Miller leased his Park and Dumbarton plantations, rented the previous year by merchant Nathan Loire, to Sam Williams and David Lamb, Natchez merchants, for $2,000 and $1,200, respectively. By contracting with freedmen to work the plantations for a share of the crops, Williams and Lamb not only collected a share of the crop as landlords but also held a lien on the cotton due the freedmen for supplies advanced. Thus, the advantages of landlord and supplier were combined, with only minimal capital investment.[44]

In another example, merchants David Singleton and David Young "united as partners" with freedmen in forty-two contracts wherein they leased lands on five plantations (Sahara, Pallowment, Novelty, Evergreen, and Morrell). The landowners received 1 bale of cotton for every 6 acres leased, the merchants furnished all supplies, and the freedmen "partners" gave Singleton and Young "a pledge and a privilege on the crop to be made for rent, and a special lien on the crop for the supplies advanced." All cotton was to be handled, shipped, and sold by Singleton and Young, with their freedmen partners receiving any surplus left after deduction of rent and supply bills.[45]

The fact that merchants were available and willing to put money into planting despite previously poor harvests may be explanation enough for their emergence as landlords. Such tenants, too, lessened the likelihood that the landowner would be forced to advance supplies himself in case freedmen failed to secure enough credit to make a crop, although he might still be forced to share his tenant's excessive costs by taking a smaller share as rent. State and local courts had generally favored the merchant's and laborer's liens over the landlord's hold on the crop for rent. Therefore, merchants became prime candidates for renting, since there was less chance that claims on the merchant-tenant's crop would eat up the landlord's rent. This judicial and legislative development (the crop lien laws) meant that merchants could also advance credits for making a crop and in turn secure lines of credit from eastern wholesalers because something of value, the crop itself, secured the debts.[46] It would be misleading, however, to assume that merchants were not securing pledges of crops before 1867. On the contrary, the crop lien laws codified what had appeared in hundreds of written agreements since 1863 and was traditionally recognized in antebellum times: factors before the war commonly advanced credits with the understanding that they would receive cotton to sell.[47] The importance of the crop lien laws goes beyond the codification in that they gave the supplier the security he needed once the planter's antebellum source of credit—slaves and valuable lands— had vanished. The fact that old-time planter Wesley

Conner was unable to secure credits in 1863, even with a mortgage on his lands, suggests how much the situation had changed.[48]

The crop lien, initially designed to aid the planter in securing credits, actually enabled the merchant to circumvent the planter and deal directly with the freedmen. Here was the security which the laborer's lien on the crops for wages had failed to provide. The crop lien laws removed all doubt as to who held the first lien on the crops in cases of dispute between landlords, tenants, laborers, and suppliers. In all cases where either the landlord or his laborers had contracted debts with the supplier for supplies advanced in making the crops, the supplier held a "prior lien" on the crops made. This lien even preceded the landlord's claim for rents from his tenants.[49]

The extent to which merchants began dealing directly with freedmen as suppliers, employers, and landlords is seen in the lien records of Adams County and Concordia Parish dating from 1868 to 1880. Dozens of examples could be listed to support the fact that Natchez merchants made the most of the opportunity, but a few will suffice to illustrate the situation. Vidalia supply merchant Isaac Friedler, one of the most successful merchants in the area, had nearly one hundred accounts with freedmen sharecroppers—mainly squads and families—on seventeen different plantations in 1874.[50] By the end of the decade, the man whom Somers might well have used as a model for his "Mr. Solomon" had increased his dealings to several hundred accounts on thirty-four plantations. His supply bill for that year amounted to over $25,000 and included mules on lease as well as lands.[51] Similarly, the forty-two accounts of merchants Singleton and Young, partners in supply, totaled $12,145 on five plantations in 1874.[52] On the other hand, to balance the picture, Mrs. Mary Rodder, a Vidalia merchant, recorded only two liens in 1869 for supplies amounting to $80 advanced to several freedmen planting crops on a portion of the Wukamore plantation.[53] The individual accounts varied from a few dollars to hundreds, and it was not unusual to find small supply merchants indebted to large supply merchants. One lien recorded in 1881 found merchant Morris Wexler advancing $100 to Vidalia supply merchant Mrs. Nannette Israel to enable her to carry on her business. In return, Mrs. Israel pledged "all her rights that she may have and now has as a furnisher of supplies to three people on Dr. Carter's River Place."[54]

One pattern that emerges from the voluminous records is the extent to which liens were often piled on top of liens. Freedmen working as sharecroppers on the plantation of D. P. Williams, an antebellum planter who had managed to hold on to his lands, agreed to ship their share of the crop to any merchants making advances by giving "a first lien over other parties who may make advances except the said Williams who shall have a like lien for any assistance afforded them."[55] This written agreement recognized the possibility of conflicting claims by various suppliers. The average lien recorded by Singleton and Young in 1874 amounted to a debt of $245 for supplies advanced, while a year later Isaac Friedler's accounts averaged less than $100. The variation in amounts supplied indicates that Friedler may have operated in many arrangements as one of those "other parties" making advances. Even those freedmen dealing with Singleton and Young probably appeared sufficiently risk-worthy to draw from "other parties," since they rented on the average over 30 acres, yet normally paid a set rent of 1 bale for every 5 or 6 acres with another 5 or 6 bales needed to pay for supplies. This accounted for one-third of the crop which 30 acres of Concordia cotton lands had yielded in 1860.

The general picture emerging from the lien records detailing the supply arrangements between freedmen, merchants, and planters is one of extreme competition, at least in the 1870s. While it is doubtful that merchants wished to advance money or supplies to freedmen heavily in debt to others, it did happen enough to show the extent of their competitiveness. More telling was the frequency with which several merchants, often as many as six, held accounts with freedmen on the same plantations. Isaac Friedler shared accounts on Helena plantation in 1873 with merchant James Pendleton, did the same on Whitehall with John Mackin, and on Shamrock with merchant J. H. Scott.[56] These competing merchants had no real territorial monopolies, although once an account was established it most likely continued to run for several years. The procedure was simple enough. Freedmen consumers of mercantile credit seldom shopped around to compare prices, haggled over interest rates, or changed creditors as a result of promotions or merchant offerings of advantages. Rather, the typical freedman shopped for the amount of advances he might obtain over the year. Merchants competed on their record or promise of advances to be had and, at least initially, by their attitude regarding supervision.[57] In time, these terms of credit became quite well established. For the merchant, such promises and accommodation were easy terms to meet, since sums and supplies advanced carried heavy interest rates and specification of crops to be planted. In the Natchez District, at least, if merchants held freedmen in a monopolistic grip, it was the hold of a *system* of competitive merchants rather than the clasp of a single country store, although the effects of such competition probably meant little to the ever-increasing debtor poverty of the customers involved.[58]

Not all planters abandoned their interest in supplying the freedmen. Some, like Jane E. B. Conner, even expanded that aspect of their enterprise. Conner had furnished the laborers on her Rifle Point plantation since 1865, continued to do so as her wage laborers became sharecroppers, and by 1880 held supply accounts with freedmen on neighboring plantations.[59] But most planters who continued to advance supplies on their own accounts to freedmen seldom moved beyond the business of their own plantations. Indeed, in 1880 only 2.5 percent of the landowners in Adams County and 8.2 percent of those in Concordia Parish were involved in merchandising enough to have such assets listed in the census book or tax records.[60] Although these figures exclude tenant-planters in the district, they do indicate that planter-landlords were not furnishing merchants in an entrepreneurial sense. Most planters who made advances stocked their stores through advances made in turn by factors and local merchants. Isaac Friedler, again, included among his numerous accounts in the 1870s and 1880s the plantation stores of five of the largest planters in the parish.[61] These planters, some of whom were probably in competition with Friedler, most likely furnished supplies as an accommodation rather than as an enterprise, in cases of emergency, or as some minimum ration in order to supplement the income of their wage hands. The fact that the "Nabob" A. V. Davis maintained a store on only one of his several plantations is typical of the landlord's interest in supply.[62] Profits, which in some instances meant the difference between solvency and ruin, were obviously made by planters with stores after 1867, but there is little evidence that more than a handful relied on the business in a large way.

It should not be assumed from what has been said above that Natchez District merchants easily adjusted to the postwar scene. For most the pace of change must have seemed maddening. Fortunes were made and lost literally overnight, and even the most established firms threw all cautions aside in the rush for profits. John Fleming and H. M. Baldwin

had been merchants in the district since 1847. They were rich and
"sober" businessmen with a sound reputation for "industry" and "honor."
In 1865, R. G. Dun's commercial credit reports estimated the firm's
capital at nearly $100,000 in merchandise and nearly $500,000 in real
estate. Within a year, however, the firm teetered on the edge of bankruptcy:

> This firm was very wealthy after the war but planting has nearly ruined them.
> They own very valuable property but it is encumbered on account of disasters.
> Their stock of goods is mortgaged. All they own except their residences are
> mortgaged. I know one claim of $25,000 they are unable to meet. And one case
> reported to me when they were unable to pay a planter a cash balance of
> $18,000 for the sales of his cotton.

Although the firm survived into the 1870s, Dun's local correspondent
was by that time unwilling to rate its creditworthiness.[63]

The firm of Meyer, Eiseman, and Company experienced similar
difficulties, but managed to avoid the worst by means of careful and
calculated reorganization plans. Dun's correspondent had confidently
rated the house as the largest in Natchez in 1866, worth over $250,000
with substantial accounts and loans outstanding. He thought of the
partners as "sober" men, somewhat "trickish," but "keen and shrewd"
traders who "would not scruple to take advantage whenever opportunity
offers." The partners had speculated heavily in cotton during the war,
amassing a substantial fortune for their efforts. In 1866, they organized
branch stores in the neighborhood and financed numerous planting en-
deavors. By the end of the year, however, the firm faced bankruptcy
and quickly moved to reorganize, turning all assets over to other members of the family. Dun's correspondent thought the maneuver somewhat "fraudulent," although he carefully noted that "their condition is
known to only a few here and not talked of at all. Those who know are
surprised and cannot understand why they should transfer property when
they have more than double enough assets to pay their debts."[64] Their
assets, however, were mainly in real estate and accounts due from
which little settlement or capital could be expected.

That such hustling for profits characterized even the most stable of
district houses indicates how much the war had changed the affairs of
business. Merchants had abandoned the business of accommodating
planters for a commission. They had become speculators and hustlers of

the most frenzied sort. For every established house that leaped into the fray, several new houses added to the growing mania. George Morgan arrived in Natchez in 1866 on the run from debts and perhaps fraud in Ohio. He set up a number of stores in the district, failing in one after the other until his death in 1870. At that time, his daughter and son-in-law reorganized the firm and continued doing "a large business with the Negro population." Their entry in the Dun ledgers noted the firm's checkered past and somewhat shaky foundation: "A sudden depression of prices will close the store. They keep matters so arranged that no money can be forced out of them by litigation. I have now in hand a claim against them where a friend of theirs left certain effects in their hands for safekeeping, cotton money, notes, etc., and I can get no statement of account from them, nor any of the effects or money."[65]

Rickey, Shelton and Company combined youth, Yankee aggressiveness, and northern capital—all to no avail. The Dun reports rated the firm as "Dishonest and Unsafe" because of its excessive expenses and its "striving to draw customers from all directions." Rufus P. Rickey was reported to have been a storeman from Jackson, Mississippi, who joined with Shelton in New Orleans to obtain backing from a Yankee capitalist identified as a former Union general. This combination was perhaps reason enough for Dun's correspondent to question the reliability of the firm. But most damaging to its reputation was the firm's willingness to contract debts on a daily basis from anyone willing to provide operating capital. The interest on these spot loans often amounted to "¼ of 1% per day or 90% per annum." Much of this capital went into cotton speculation as well as supplies for the provisions trade and plantation business. In January 1867, the partners declared bankruptcy and left behind them some thirteen suits for individual amounts exceeding $28,000.[66]

Another Northerner, Lewis Frager, settled in Concordia and amassed a substantial fortune with his cotton business and supply arrangements. By 1869, he owned three plantations. Then his affairs soured. Losing money on cotton and unable to meet his obligations to his eastern wholesalers, Frager looked to his brother in Louisville and his father in New York for assistance. But by 1872 he had been forced to sell everything and leave the district.[67] Thomas J. Pollock, a well-established merchant in the district since the 1840s, lost his business after the war. He remained at the store as a salaried manager.[68] The once wealthy planter Samuel L. Winston tried his hand at merchandising after the fighting had stopped. Dun's correspondent ranked him in 1858 as a "gentleman of respectability" who "sometimes sports a title." By 1867, the record listed him as one of those storemen "not worth a cent," who "never was known to pay a debt if he could avoid it."[69]

The list of such difficulties could be extended, but enough has been noted to make the point. The business of provisions and trade in the immediate postwar years was fraught with risk and uncertainty. The nature of the business had dramatically changed. In the first two years after the war, the temptation to turn from trade to planting and speculation in cotton-buying overwhelmed district merchants. Few survived the brief period of speculation and overexpansion. Those who did settled down to the business of providing plantation supplies to their freedmen tenants, and by the mid-1870s some slight order was beginning to emerge out of the earlier chaos. Houses like Fleming and Baldwin continued their interest in planting as well as supply, often owning plantations outright. But the most stable of the surviving firms leased plantations, supplied freedmen and planters, and, if the Dun reports are a reliable basis for judgment, shied away from the business of cotton speculations of the kind common during and shortly after the war. The mania for cotton planting and the willingness to risk great amounts of capital in support of any and all who would make a crop had ended almost as quickly as it had begun.

THE SURVIVAL OF THE ANTEBELLUM PLANTER

A good many antebellum planters managed to hold on to their lands by relying on local merchants for credits or on their services as agents. William Conner's family, for example, finally settled in 1892 the mortgage on their Hayhayan plantation, given in 1866 to a coastal factor and a local merchant for supplies.[70] But the question that arises is whether the antebellum planter elite endured as a group or class of landowners in the generation after the war. The bulk of the literature, much of it fictional, suggests that the antebellum planter class was soon displaced by a merchant class that had used the crop lien legislation to gain control over the planter's lands, laborers, and crops. Recently, however, Professor Jonathan M. Wiener has found a significant persistence of antebellum planters in Alabama as a class of landowners and economic elites. Merchants in Alabama were relegated to dominance in the nonplantation hill country, with planters remaining supreme in the black belt regions. Briefly, Wiener believes that the planter class in Alabama acted as a self-conscious class in its fight to undermine the merchant's claim to the tenant's crops for advances made. The means for defeating its bourgeois opponents was the law of 1871 which gave the landlord a lien "superior to all other liens, for rent and advances made" to tenants. This law forced merchants to concentrate on supplying the white, yeoman landowners of the hill country, while the planters dominated the supply situation in the black belt area.[71]

The data for the Natchez District, although employing a different measurement technique from Wiener's, only partly support his conclusions. Table 9 below is based upon the 1860 and 1870 manuscript census schedules for Adams County as well as other local records, including wills and mortgage documents. It shows that less than 25 percent of the county's antebellum planter families survived as holders of plantation estates in 1870. More may have survived but remain hidden because of the nature of the records. Of those families surviving, their total holdings also survived basically intact as plantation estates, although 15 percent fewer acres appeared in the relevant improved acreage records in the census.[72]

When studied in categories of size defined by acres improved, we see that the largest estates experienced the greatest decline in land under cultivation as well as a greater incidence of loss. The smallest holdings in 1860 experienced a 140-percent gain in acreage on the average, though even here the typical family lost acreage. Nearly one out of three planter families survived the war as landowners, but most of the survivors lost over half the acreage they had cultivated in 1860. Those who gained, some 29 percent of the surviving families, gained significantly, with some nearly doubling their 1860 acreage improved.[73]

When the materials for Adams County are viewed according to the range of acreage owned, the evidence suggests that the largest group of survivors from 1860 owned medium-sized plantations in 1870. Yet, those survivors who owned the largest plantations in that year equaled 41 percent of the number of planters who owned similar estates before the war. (See Table 10.) By 1880, according to the listings of estates and planters in the manuscript census, the number of surviving planters had fallen to 21 percent of the antebellum group. Again, the largest

Table 9
Persistence of Antebellum Planter Families, Adams County, 1860-1870

Percent of planter families in 1860 owning plantations in 1870			24

Acres Improved in 1860:	1,000 acres	999-500	499-100
Percent of surviving families losing land	78	60	57
Average amount of 1860 holdings lost (pct.)	50	67	36
Median loss of 1860 acreage (pct.)	65	67	29
Percent of surviving families gaining land	22	30	35
Average amount of 1860 holdings gained (pct.)	92	41	140
Median gain (pct.)	68	25	58
Percent of surviving families neither losing nor gaining land	0	10	8

SOURCE: Manuscript Census, Adams County, 1860, 1880; Manuscript Tax Rolls, Adams County. 1860-1870; Probate Records, Adams County, 1870.

Table 10
Range of Acres Owned by Antebellum Planters in Adams County Who Survived as Planters in 1870 and 1880

Acres improved	− 100	100 +	200 +	500 +	700 +	1,000 +
1870 (pct.)	8	14	31	16	16	15
1880 (pct.)	9	9	39	19	15	9
1860 (pct.)	18	18	29	14	13	7
Survivors in 1870 as a percent of 1860 planters	11	18	22	26	29	41
Survivors in 1880 as a percent of 1860 planters	11	10	28	29	25	24

SOURCES: Manuscript Census, Adams County, 1860, 1870, 1880; Manuscript Tax Rolls, Adams County, 1861; Liens and Mortgage Records, Adams County, 1860-1880; Probate Records, Adams County, 1870-1880.

group of survivors owned medium-sized plantations. Those who owned the largest estates represented only half the group in 1870 and thus less than a quarter of the antebellum group of large planters.[74]

The data for Concordia, while more difficult to compile and less precise, are equally revealing. The censuses for 1870 and 1880 are nearly useless for our analysis, except in a corroborative sense. In 1870, the census enumerator listed the operators of estates and not the owners. In 1880, the census shows sharecropping tracts within the plantations but not the total estate as a single unit, thus giving the misleading notion that the old plantations had disappeared. More helpful are the surviving tax records for 1881 which specify owners and total acres. Unfortunately, these records lump together acres improved with acres unimproved, making it impossible to determine comparative acreage cultivated. Nevertheless, by taking the tax records, the manuscript census tracts, along with other local documents ranging from courthouse records to plantation manuscripts, we can identify seventy of the 134 antebellum plantations as to owners, names of estates, acreage owned, and approximate locations for both 1860 and 1881. Of these seventy, as Table 11 indicates, 54 percent remained in the same family a generation after the Civil War. As in the case of Adams County a decade earlier, the persistence was accompanied by a 19-percent loss of total acreage. The bulk of these surviving plantations were large estates, but the typical place had lost more than one-fourth of its antebellum acreage. Those gaining, though fewer, enjoyed a similar percentage increase in size, also—one-fourth.[75]

Who obtained the lands lost by antebellum planters? In Concordia, in 1860, the census records listed 134 farms over 200 acres. By 1880, the list had increased to 253 farms. Or, to put it another way, there were 208 separately owned farms in the parish in 1860 compared to 359 in 1880. Many of these lands were subdivided in death, broken apart and divided to settle accounts, and confiscated for bad debts. But few plantation estates disappeared. Rather, they grew smaller while retaining their core size and character. The typical family losing lands had sold off small chunks of marginal lands while not touching the main body.

A significant number of the lost estates were claimed by members of the antebellum planting class. Members of the wealthy Surget family, with cousins sprinkled liberally among the district's leading antebellum clans, increased their holdings substantially. The bulk of the lost estates

Table 11
Persistence of Antebellum Planter Families, Concordia Parish, 1860-1882

	1,000 acres	999-500	499-100
Percent of Plantations in 1860 in same family in 1881			54
Acres improved and unimproved in 1860:	1,000 acres	999-500	499-100
Percent of surviving plantations losing land	52	50	40
Average amount of 1860 acreage lost (pct.)	27	48	29
Median loss (pct.)	31	40	29
Percent of surviving plantations gaining land	32	50	0
Average amount of 1860 acreage gained (pct.)	50	33	0
Median gain (pct.)	50	26	0
Percent of surviving plantations neither losing nor gaining land	16	0	60

SOURCES: Manuscript Census, Concordia Parish, 1860, 1870, 1880; Manuscript Tax Rolls, Concordia Parish, 1861-1896; Liens and Mortgage Records, Concordia Parish, 1860-1890.

and divided parcels were obtained by new names in the records, although it is impossible to know for certain how many were actually newcomers rather than previously unidentified cousins and in-laws. Even the record of land sales fail to yield genealogical certainty.[76]

Some of these lands ended up in the hands of those successful merchants who held claims for advances made. Merchant John Fleming sued William Sojourner in 1875 for title to 1,317 acres mortgaged to him for supplies. In most cases, however, the merchants preferred to sell the lands for cash or in some arrangements wherein the buyer would agree, as part of the sale, to do business with the merchant-realtor.[77] A close reading of the manuscript census, extant tax rolls, and numerous deed and mortgage records leaves the impression that only a small percentage of district merchants owned plantations in 1880. That so few owned land is hardly surprising, since planters had mainly given their mortgages to coastal factors and eastern firms rather than to local merchants. In addition, few merchants, operating on credits from eastern wholesalers, had the capital resources to buy lands even at cheap prices. And the chances for success were never so great as to insure fortunes for all. In fact, only 14 percent of the merchants in 1870 had been in the furnishing business in 1860, and only 27 percent of the 1870 group were doing business in 1880. This high turnover meant that at any one time in the 1870s the average merchant was either on the verge of dropping out of business or just getting started.[78] Merchants were able to secure as much control over the freedman's crop by renting and supplying as by direct ownership of the land.

Whether these data suggest a remarkable persistence of the district's plantation elite is a matter of perspective. For every planter family surviving in 1870 in Adams County, two or three had disappeared. For every plantation in our Concordia sample that remained in the same family, another had been sold, lost or willed away. In addition, most of the survivors had lost significant parts of their estates. Even less clear from the records is how these surviving planters had survived. Owning the land in 1880 did not make one a planter in the antebellum sense of that word. More and more of the surviving planters found themselves operating as absentee landlords with their plantations leased to a local merchant and, in a few cases, even to groups of freedmen. Many of the largest of these survivors had been absentee planters before the war, but then their managers had been salaried employees operating according to the dictates of the landlord. By 1880, these same plantations were being leased to men on far different terms of accountability. These new managers were seldom the agents of the surviving planting class as much as they were storemen and entrepreneurs doing business on their own account with their own production schedules and according to their own managerial decisions and goals. The A. V. Davis family, for example, continued after the war to live in New Orleans as absentee owners of numerous plantations in the delta by leasing their lands to local merchants and white tenants.[79] James Surget leased his Ashley plantation to merchants Karl Kehmand and Berthold Lehman in 1881 for $3,800 payable to his agent, Isaac Friedler.[80] Erusture Surget had been leasing another of their holdings for years to merchant Christian Swartz for $2,000 and taxes.[81] And Mrs. Jane E. B. Conner managed to live in Saint Louis for a generation after the war partly off the income received from leasing her Concordia plantation to several merchants in the Natchez District.[82]

CONCLUSION

By 1880, Natchez District merchants had replaced planter-landlords as the freedman's chief source of supply. In most cases, they accepted whatever system of land tenure was in existence on the plantation. Merchants supplied freedmen on a share wage basis (a share of the crop paid as wages), on a share rent basis (where freedmen rented lands paying a share of the crop as rent), and on a fixed wage basis (where freedmen worked as wage hands with their wages paid at the end of the year). In all instances, the merchants vigorously competed with one another for customers among the freedmen. But within a few years the mere presence and functioning of the merchant tended to institutionalize the system of sharecropping over fixed wages as the dominant form of land tenure in the area. Essentially, merchants found little reason, either in terms of efficiency or social control, to assist the return of gang labor conditions or set wages. First, a return to the gang labor conditions of slavery (remembering now that an essential feature of sharecropping was the single-family working arrangement) meant costly supervision which the merchant could neither afford nor profit from. Second, fixed wages made even less sense to the merchant as the price of cotton began its downward spiral after 1870. Pegging wages to the harvest freed the merchant from committing himself to a wage due that might be above the actual returns from the crop sale. Finally, neither the planter-landlord nor the freedman could easily reinstitute fixed wages or gang labor conditions once they became dependent upon the merchant for supplies in making the crop. At this critical juncture in the system's development, merchants held liens on supplies advanced that superseded both the planter's claim on the freedman's crop for rent and the freedman's claim on the planter's crop for wages due. With the merchant in control of the crop, the only way fixed wages, gang labor, or any other alternative to sharecropping might have emerged from the merchant's standpoint is if the system had failed to produce a crop. But even a cursory reading of the production statistics for the district in 1880 clearly indicates the extent to which sharecropping met the merchant's fondest expectations.

It was the accommodationist character of the merchant that enabled him to fit into the postwar situation, first as a petty entrepreneur and finally as a dominant supplier-landlord. The merchants did not originate sharecropping, but they did accommodate themselves to it as a system of supply, land tenure, and class relations. Its institutionalization generally reflected the merchant's rational accommodation to a system originating in a context of complex historical realities ranging from racism to the much noted "traditionalism" of the region's "plantation mode of production."

For the antebellum planter class, the merchant's role as chief source of supply was both resented and welcomed. Planters who had lost lands to their merchant creditors undoubtedly saw the merchant's role as the key to their own decline. For others, the merchant himself became a trusted tenant and agent willing to deal with the "faithless" laborers on terms "too degrading to even consider." Some saw the merchant's willingness and enterprise as the very cause of the freedman's so-called demoralization. One thing is certain, however: few surviving antebellum planters, even though they continued to own plantation estates, were planters in the antebellum sense. For some, especially those antebellum absentee owners, the new era meant replacing the overseer-manager with the merchant-manager. But for others, especially the antebellum resident planters, survival as landowners usually required giving up their resident planter status. A great many, like Mrs. Catharine Davis, T. Casey Witherspoon, and Mrs. Jane E. B. Conner, left the neighborhood to live in St. Louis, New Orleans, Memphis and even in Philadelphia, with their plantations leased out and managed by others.[83] When the antebellum planter William F. Miller agreed to lease his Park and Dumbarton plantations to Natchez merchants in 1869, he fully understood that his tenants would then contract with freedmen to work the plantations as sharecroppers and that neither he nor the merchants would live on the places. He also understood that such a deal gave the merchants a lien on the cotton due to the freedmen for supplies advanced and the right to a share of the freedman's crop as rent. Therefore, the merchants had the advantages of landlord and supplier with but a minimum capital investment and without actually owning the land. Miller had retained his ownership of the plantations, but he was no longer a planter.[84]

NOTES

1. Robert Somers, *The Southern States Since the War, 1870-1871* (University, Ala.: University of Alabama Press, 1965), pp. 241-46. First published in 1871.

2. Samuel Postlethwaite to J. A. Gillespie, February 22, 1874, J. A. Gillespie Papers, Louisiana State University, Baton Rouge, Louisiana. Postlethwaite

confided to Gillespie that he intended to try share-leasing in 1874 and that he planned to keep the freedmen a little in debt for supplies and animals, since it would "make it easier to contract for another year."

3. G. D. Reynolds to W. E. Strong, March 25, 1865, Records of the Bureau of Refugees, Freedmen, and Abandoned Lands (hereinafter cited as BRFAL), Record Group 105.

4. Shields to Mercer, August 7, 1866, William Newton Mercer Papers, Louisiana State University, Baton Rouge, Louisiana; Gillespie to an unknown officer of the Freedmen's Bureau in Natchez, May 29, 1865, Gillespie Papers.

5. G. D. Reynolds to S. Thomas, August 10, 1965, BRFAL, Record Group 105.

6. Ibid., Captain B. B. Brown, U.S. Army, to G. D. Reynolds.

7. A good many freedmen in the area collected substantial sums of money for their army service. One white in the district noted that "two days ago a regiment of twelve hundred men were disbanded at this place, where it was paid off in the sum of $40,000, but at this moment very few have a cent left." King to Lin, May 17, 1865, Thomas Butler King Papers, Southern Historical Collection, University of North Carolina, Chapel Hill, North Carolina.

8. G. D. Reynolds to Assistants, June 1, 1865, BRFAL, Record Group 105; see also U.S. Congress, House, Circular Issued by General O. O. Howard, No. 11, July 12, 1865, 39th Cong., 1st Sess., 1865, published in House Exec. Doc. 70: 185-86.

9. J. D. Shields Papers, Louisiana State University, Baton Rouge, Louisiana.

10. Shields to Buckner and Newman, July 10, 1866, ibid.

11. Shields to J. R. Jones, September 20, 1866, ibid.

12. Shields to Jones, October 10, 1866; Shields to Buckner, November 23, 1866, ibid.

13. Shields to Buckner, November 23, 1866, ibid.

14. P. Shawn Percy to Shields, January 10, 1867, ibid.

15. Shields to Percy, August 17, 1867, ibid.

16. See Liens and Mortgage Records, November 12, 1865, Concordia Parish, Louisiana, Office of Records, Vidalia, Louisiana.

17. December 16, 1865, ibid.

18. See Postlethwaite to Gillespie, January 15, 1871, Gillespie Papers; Somers, *Southern States Since the War*, pp. 241-46. Prior to the war, a planter's word and reputation secured him the advances needed to make the crop largely because of his control over labor. Planters who had made and delivered the crop year after year were seldom required to mortgage either their land or slaves for the credits advanced. But slavery's end reduced the planter's control over his labor to the point where a supplier could not be sure about the crop's delivery. More importantly, slaves were no longer a capital asset.

19. Liens and Mortgage Records, September 7, 1865, December 10, 1865, December 15, 1865, January 14, 1866, Concordia Parish, Louisiana, Office of Records, Vidalia, Louisiana.

20. March 14, 1866, ibid.

21. Liens and Mortgage Records, February 26, 1866, Adams County, Mississippi, Office of Records, Natchez, Mississippi.

22. March 5, 1866, ibid.

23. May 2, 1866, ibid.

24. June 5, 1866, September 29, 1866, ibid.

25. Buckner, Newman & Co., V. John L. Metcalf et al., Chancery Records, Adams County, Mississippi, Book D, 119 (1869).

26. See the letters of F. Lloyd King, King Papers.

27. F. L. King to Mallory King, January 18, 1866, ibid.

28. F. L. King to Lin, June 8, 1866, ibid.

29. February 13, 1867, ibid.; Liens and Mortgage Records, August 6, 1869, Concordia Parish, Louisiana, Office of Records, Vidalia, Louisiana. Surget had worked the plantation himself with sharecroppers in 1868 by giving a mortgage to the banking firm of Britton-Knotz in order to secure the capital needed to provision the place.

30. *Natchez Democrat*, November 18, 1867, p. 1, December 23, 1865, p. 1. In this type of arrangement, a merchant could easily end up owning the very lands he had sold on commission as a real estate agent.

31. E. Jeffords Accounts, 1866, BRFAL, Record Group 105.

32. February 1864, Records of the Adjutant General's Office, Colored Troops Division, Record Group 366, National Archives, Washington, D.C.

33. June 1, 1864, September 26, 1864, ibid.

34. R. S. Hart to William P. Mellon, February 1, 1864, ibid.

35. Hart to Mellon, January 26, 1864, ibid.

36. Hart to Mellon, June 24, 1864, ibid.

37. Mellon to S. Richardson, October 10, 1864, April 29, 1865, ibid.

38. Liens and Mortgage Records, October 3, 1865, Adams County, Mississippi, Office of Records, Natchez, Mississippi.

39. See Merle C. Nutt, *The Nutt Family Through the Years, 1635-1973* (Phoenix, Ariz.: Merle C. Nutt, 1973), pp. 93-137.

40. See Liens and Mortgage Records, April 22, 1870, Adams County, Mississippi, Office of Records, Natchez, Mississippi.

41. January 1, 1870, ibid.

42. March 19, 1870, ibid.

43. Postlethwaite to J. A. Gillespie, January 15, 1868, Gillespie Papers.

44. See Liens and Mortgage Records, November 9, 1869, Concordia Parish, Louisiana, Office of Records, Vidalia, Louisiana.

45. January 24, 1874, ibid.

46. Mississippi, *Laws of Mississippi* (1867), Sec. 1, 2, 7. The most important parts of the lien laws in regard to our study were:

    (1) All debts contracted for advance of money for purchase of supplies,
farming utensils, working stock, or other things necessary for the
cultivation of a farm or plantation, shall constitute a prior lien upon
the crop and also on the animals and implements employed to cultivate the crop which shall have been purchased with the money so
advanced.

    (2) That when any owner or lessee of any plantation or farm, shall make
any contract for a share or shares of the crop, in lieu of wages, and
said owner or lessee shall make advances of money, provisions, or
clothing, such owners or lessees shall have a lien on the share of
such laborers' crop for payment of the same.

    (3) That it shall be lawful to convey by way of mortgage or deed of
trust, any crop of cotton, corn, or agricultural product, now being
produced, or to be produced within fifteen months from the date of
such mortgage: provided that nothing shall interfere with any prior
lien granted by the provisions of this act for supplies and means
furnished to grow the crop.

47. See Harold D. Woodman, *King Cotton and His Retainers: Financing
and Marketing the Cotton Crop of the South, 1800-1925* (Lexington, Ky.:
University of Kentucky Press, 1968), p. 296.

48. Liens and Mortgage Records, April 30, 1866, Concordia Parish, Louisiana, Office of Records, Vidalia, Louisiana.

49. See Somers, *Southern States Since the War*, pp. 241-46; *Natchez Democrat*, December 7, 1870, p. 1. Although the primacy of the merchant's claim
over the landlord's claim on the crop was not settled by the courts until the
1870s, it is clear that merchants in the district were able to secure a first claim
on the crops as a privilege shortly after the passage of the lien laws. The
merchant's success in this matter followed upon his emergence as something of
a middle man between the freedman and his landlord. Planters commonly
supplied their croppers by sending them to the local merchant. In other instances, the merchant acted as the landlord's agent in renting and supervising the
plantation. In some cases, the landlord was forced to recognize the merchant's
claim for supplies over his claim for rents or face the possibility of losing the
merchant's services for the following year.

50. Liens and Mortgage Records, August 13, 1874, Concordia Parish, Louisiana, Office of Records, Vidalia, Louisiana.

51. August 11, 1880, ibid.

52. January 24, 1874, ibid.

53. March 19, 1869, ibid.

54. May 1, 1881, ibid.

55. Liens and Mortgage Records, February 18, 1870, Adams County, Mississippi, Office of Records, Natchez, Mississippi.

56. Liens and Mortgage Records, April 19, 1874, June 1, 1874, August 13, 1874, Concordia Parish, Louisiana, Office of Records, Vidalia, Louisiana.

57. March 19, 1869, ibid.

58. This point is different from that expressed by Roger L. Ransom and Richard Sutch's suggestion that southern country stores enjoyed a distinct territorial monopoly, based upon market distance. According to their findings, an average distance of 5.5 to 9.0 miles between stores represented a serious barrier to competition. Although it is doubtful that such distances were too much for a monthly shopping trip in the nineteenth-century South, since it could easily be walked in a day's time, the actual sharing of single plantation accounts by merchants suggests how little their thesis applies to the Natchez District. More importantly, Ransom and Sutch's analysis is based on area divided by the number of stores and not on population patterns or actual store locations. In Concordia Parish in the 1870s, merchants were located at fifteen different geographic points (Vidalia, twenty-three stores; Monterey Landing, two stores; Black Hawk Landing, three stores; Gibsons Landing, two stores; Waterproof P.O., one store; Bulletts Bayou, three stores; Normandy Landing, two stores; Union Point, one store; Fairview Landing, five stores; Concordia Lake, one store; Lake Saint John, two stores; Frog Moor, one store; Good Hope Landing, two stores, Bowern Point, one store; and Tooleys P.O., one store). The spread of these stores, not to mention those not listed in the Dun Records, placed numerous stores in competing distance of one another. In Adams County, the bulk of the stores were located in Natchez. Freedmen regularly traveled to Vidalia and Natchez for supplies, or, as was most common, storemen traveled to the plantations with goods for their more localized commissaries. See R. G. Dun & Co. Collection, Baker Library, Harvard University Graduate School of Business Administration; Roger L. Ransom and Richard Sutch, *One Kind of Freedom: The Economic Consequences of Emancipation* (Cambridge, England: Cambridge University Press, 1977), pp. 126-48.

59. Liens and Mortgage Records, February 18, 1875, September 7, 1880, January 20, 1881, Concordia Parish, Louisiana, Office of Records, Vidalia, Louisiana.

60. U.S. Census (1880), Manuscript Population and Agricultural Schedules, Adams County, Mississippi, and Concordia Parish, Louisiana; Manuscript Tax Rolls (1861-1869), Concordia Parish, Louisiana, Office of Records, Vidalia, Louisiana.

61. Liens and Mortgage Records, August 13, 1874, June 1, 1876, August 11, 1894, Concordia Parish, Louisiana, Office of Records, Vidalia, Louisiana.

62. Manuscript Tax Rolls (1871), Concordia Parish, Louisiana. Many more planter-landlords would have maintained stores had they been able to secure the capital for supplies. Samuel Postlethwaite, for instance, hoped to secure a loan in 1871 and thus be "independent of all merchants." But much to his regret only local merchants would loan money to him for a pledge of the crops. All others, meaning coastal factors, wanted both a lien on the crops and a mortgage on his lands. In the end, Postlethwaite gave up on the idea of a store on the place and instead took a smaller share of the crop from the freedmen in order that they might deal directly with local merchants. See Postlethwaite to Gillespie, January 15, 1871, October 1, 1874, Gillespie Papers.

63. Mississippi, Adams County, Fleming and Baldwin, 1847-1872, R. G. Dun & Co., Collection, Baker Library, Harvard University Graduate School of Business Administration.

64. Ibid., Meyer, Eisman & Co., 1866-1867.

65. Ibid., Morgan & Co., 1866-1870.

66. Ibid., Rickey, Shelton & Co., 1866-1867.

67. Ibid., Louisiana, Concordia Parish, Lewis Frager, 1868-1872. Frager later returned to the district as a wealthy landowner.

68. Ibid., Mississippi, Adams County, Thomas J. Pollock, 1847-1870.

69. Ibid., Sam'l L. Winston, 1857-1870.

70. Liens and Mortgage Records, March 14, 1866, January 3, 1892, Concordia Parish, Louisiana, Office of Records, Vidalia, Louisiana.

71. See Jonathan M. Wiener, "Planter Merchant Conflict in Reconstruction Alabama," *Past and Present* 68 (August 1975), pp. 73-94.

72. Lien and Mortgage Records, 1865-1875, Adams County, Mississippi, Office of Records, Natchez, Mississippi; Probate Records, 1870-1880, Mississippi, Office of Records, Natchez, Mississippi; U.S. Census (1860), Manuscript Population, Agriculture, and Slave Schedules, Adams County, Mississippi; U.S. Census (1870), Manuscript Population and Agricultural Schedules, Adams County, Mississippi; U.S. Census (1870), Manuscript Population and Agricultural Schedules, Adams County, Mississippi. The definition of planter includes the immediate family as well as such members as could be determined.

73. Ibid.

74. Ibid.

75. Liens and Mortgage Records, 1860-1890, Concordia Parish, Louisiana, Office of Records, Vidalia, Louisiana; Manuscript Tax Rolls (1861-1896), Concordia Parish, Louisiana, Office of Records, Vidalia, Louisiana; Police Jury Minutes, October 5, 1885, Concordia Parish, Louisiana, Office of Records, Vidalia, Louisiana; U.S. Census (1860), Manuscript Population, Agriculture, and Slave Schedules, Concordia Parish, Louisiana; U.S. Census (1870 and 1880), Manuscript Population and Agricultural Schedules, Concordia Par-

ish, Louisiana. See also Roger Shugg, *Origins of Class Struggle in Louisiana, 1840-1875* (Baton Rouge, La.: Louisiana State University Press, 1939, 1968), pp. 234-73.

76. Liens and Mortgage Records, 1866-1885, Adams County, Mississippi, Office of Records, Natchez, Mississippi.

77. Fleming and Baldwin *v.* William Sojourner, Chancery Records, Adams County, Mississippi, Book O, p. 142 (1875).

78. The above analysis is based upon a careful examination of the manuscript census records, tax rolls, lien and mortgage records, court documents, plantation papers, and newspapers covering the years 1850 to 1890 for the Natchez District.

79. Mississippi, Adams County, A. V. Davis, 1859-1872, R. G. Dun & Co. Collection.

80. Liens and Mortgage Records, February 2, 1881, Concordia Parish, Louisiana, Office of Records, Vidalia, Louisiana.

81. Liens and Mortgage Records, January 1, 1871, Adams County, Mississippi, Office of Records, Natchez, Mississippi.

82. Liens and Mortgage Records, June 1, 1887, Concordia Parish, Louisiana, Office of Records, Vidalia, Louisiana.

83. Ibid.

84. Liens and Mortgage Records, November 9, 1869, ibid.

• 6 •

THE FREEDMAN'S RELATIVE EFFICIENCY

In 1860, William Leum owned a medium-sized plantation in Concordia Parish: sixty-two slaves and 430 improved acres. Although prosperous enough, Leum's cotton harvest in 1859 fell seriously below the per worker yields of neighboring plantations working the same general number of slaves. Whereas Leum shipped 4.2 bales for each slave (ages ten to sixty-five), his typical counterpart sent 6.9 bales to market. The reasons for Leum's below-average per worker yields may have included managerial decisions about crop emphasis or the low efficiency of his group of slaves, but the fact that his improved acreage was valued at just half that of the lands in his immediate neighborhood suggests more obvious factors. Since his plantation was located in a stretch of delta interlaced with canebreaks and swamps, Leum's low per worker productivity may have been a consequence of farming lands susceptible to flooding and the often disastrous seepage common to the lowlands. Even best of managers might end the season in the delta with poor yields caused by a weak levee or early and persistent rains.[1]

The productivity of Leum's workers in 1859 is of interest here because his was one of the plantations still in the hands of its antebellum owner in 1880. Leum not only survived the war, but he also held on to most of what he had accumulated under slavery. Some adjustments had occurred, but overall the old planter had managed quite well for himself in view of the fact that many of his more productive antebellum neighbors were nowhere to be found in the census of 1879. Having reduced his holdings of unimproved acres from 1,830 acres to 1,150 acres, Leum expanded his improved holdings from 430 acres to 665 acres by

1880, although only 421 acres were farmed in 1879. His work force, now sharecroppers, had increased by ten from forty-one (ages ten to eighty-five) to fifty-one hands. More striking even than Leum's survival and enterprise is the fact that he bettered his antebellum returns of cotton per worker. By working sharecroppers on family patches, Leum produced 7.1 bales of cotton per worker with an investment of $22.68 in tools and implements per hand in comparison to his 4.2 bales and $21.27 invested in 1859. Does this mean that Leum's sharecroppers worked at higher levels of personal efficiency than reached by his slaves in 1859? Possibly, but when Leum's corn production is considered— 3.9 bushels per worker in 1879 in contrast to 42.2 bushels per worker in 1859—it seems clear that the freedman's relatively high cotton yields were achieved only through a drastic reduction of acreage devoted to corn.[2]

Another planter, Leum's neighbor, A. J. Welch, failed to survive the war as a plantation owner. His relative youth in 1860, forty-seven years of age, suggests that Welch probably sold or lost his place while still alive. In any case, his Grand Cut Off plantation was large enough in 1860, 700 improved acres and 148 slaves, to place him well into the district's upper-middle planter category. Although his farm's productivity in cotton per worker fared poorly in comparison to the average for his size group, it seems to have resulted from something other than unfavorable locations or inferior soil conditions. Rather, the cash value of Welch's improved acreage exceeded the average per acre value of those Concordia planters who owned a similar number of slaves. But Welch emphasized corn to a much greater degree than average and tended to work fewer acres per hand than was normal. As a result, his cotton crop suffered in comparison.[3]

Welch's disappearance from the scene hardly affected the durability of his plantation as a unit. On the contrary, its postbellum owner, W. L. Shaw, had actually increased the plantation's size from 1,700 unimproved acres to 2,000 acres, from 700 improved acres to 1,020 acres, and from 86 workers to 150, while more than doubling its cotton production from 406 bales to 1,015 (400-pound bales). Shaw worked the place with sharecroppers and managed to exceed the per worker cotton production achieved by Welch with slaves—6.7 in comparison to 4.7 bales. At the same time, he spent less on tools and implements—$45.46 compared to $162.79. He employed fewer animals per hand—0.8 compared to 0.9—and he farmed fewer acres improved per hand—6.8 compared to 8.1. But here again it is clear that Shaw's ability to produce relatively more cotton per worker with sharecroppers than Welch had with slaves resulted largely from his emphasis on cotton. Whereas Welch's enterprise yielded 6,000 bushels of corn in 1859, Shaw harvested only 1,815 bushels with nearly twice the labor force.[4]

A third plantation in the immediate vicinity, John Miller's thirty-eight slave Point Breeze, also survived the war intact but in the hands of another. Lewis Frager had purchased Miller's place in the 1870s along with two others (Union Point and the Red River Place), while renting a third (DeLoche). Point Breeze was the smallest of the group, having only 250 improved acres in 1880. Comparatively speaking, Frager's twenty-three adult croppers and the same number of hired wage hands outproduced Miller on a per worker basis in cotton by an entire bale, 5.3 to 4.3. At the same time, the amount of money invested per worker in tools and implements on the place declined from $107.69 in 1859 to 80 cents in 1879. As with the others, however, Frager's hands and croppers harvested only 4.3 bushels of corn in contrast to Miller's 30.7 bushels of corn per worker or slave.[5]

Alex Campbell, a newcomer with no visible roots in the area, worked a large force of hired hands—forty-two adults—and leased 268 acres of land to eight families of sharecroppers in 1880. When viewed according to status, Campbell's croppers outproduced the wage hands 9.5 bales to 8.5 bales of cotton per hand and 22 bushels of corn to 14 bushels, and farmed fewer acres—11.2 improved per hand compared to 16.2. Taken together as one plantation, Campbell's croppers and hands produced 8.9 bales of cotton per worker in comparison to the 6.9 bales produced on the average by antebellum planters of the same size. They invested only $53.35 in tools and implements in comparison to the $94.48 spent in 1859. With regard to relative performance, however, Campbell's cotton crop was achieved at the sake of corn, as he harvested only 1,030 bushels in contrast to the 3,219 produced on the average slave plantation working a similar number of hands in Concordia Parish in 1859.[6]

Across the river in Adams County, Thomas H. McLowen produced far less cotton per worker with his sharecroppers than had similar sized antebellum operations. Thirteen groups of sharecroppers turned out only 1.7 bales per hand in comparison to the typical yield of 3.4 bales on slave plantations of similar size in the area. But McLowen's cotton harvest was countered by extraordinarily high corn returns of 133.4 bushels per hand in contrast to the 45.6 bushels produced in 1859.[7] Close by to McLowen's place, on the other hand, ten families of share-croppers managed to harvest only slightly below the average slave's production. Specifically, these croppers made 3.1 bales and 31.9 bush-els of corn while employing tools and implements costing only $20.10 per hand. This investment was far below slavery's average of $56.47 for the district as a whole.[8]

If nothing else these sketches make one thing quite clear: the much argued thesis that the end of slavery had undermined the freedman's personal efficiency in cotton farming is hardly conclusive.[9] In almost every case cited above, the freedman's personal productivity in cotton exceeded the per worker production levels achieved in slavery in 1859. However, the freedman's productivity in cotton was not duplicated in corn or in any of the other crops produced in the years before the Civil War. It is therefore reasonable to suggest that the freedman's productiv-ity reflected mainly the farmer's concentration on cotton as a cash crop. Corn was deemphasized in contrast to before the war when the average slaveholder devoted almost as many acres to the one as to the other.

METHODOLOGY

But these illustrations, few as they are, are equally inconclusive. For more compelling evidence, aggregate data, of course, and a systematic approach are needed. Unfortunately, working with the data is a diffi-cult task. The manuscript census, for example, although the prime source of the data required, is extremely misleading unless used with the utmost caution. Not only must the idiosyncrasies of the various enumerators be recognized, but also the seemingly endless task of tran-scribing dozens of bits of information about thousands of forgotten individuals from the manuscripts to code sheets is a duty that lends itself to the simplest, yet most devastating, of errors.[10] Since solving the problem depends on pinpointing the production schedules of the numer-ous farmers in the district in both 1859 and 1879, several important decisions must be made regarding the limitations posed by the data's arrangement and content. We can be fairly certain about what was produced by whom in both 1859 and 1879, since each farmer's produc-tion was listed alongside his name in the manuscript census. It is more difficult to determine inputs, however. It is almost impossible to determine whether freedmen worked with better or poorer quality tools than the slaves had, since the manuscript census, while listing the values of tools owned, failed to list the value of tools employed. As a result, many freedmen croppers and renters, although supplied tools by landlords and merchants, frequently appear in the census as having had no or fewer tools on their farms, and no reliable method is known for linking up the croppers with their suppliers of tools.[11] Along these same lines, the amounts listed as the value of tools usually included the value of equipment, which meant that the estimated value of mills, presses, and gins was always part of the total sum listed for each farmer in both periods. But since few freedmen owned much beyond their hoes and rakes in 1879, any attempt to compare the value of tools utilized per worker would seriously overestimate the value for slaves in 1859 and for most white owner-operators in the postwar period.

Even more difficult to resolve is the question of labor. In slavery, the census enumerator simply listed the sex and age of each individual slave. It is thus relatively easy to misjudge the personal productivity of the individual slave inasmuch as domestic servants were included as field hands. It is also easy to misjudge the participation rate and effect of adolescents. But freedom forced the enumerator to deal with the question of hired labor in a fashion that is far more misleading, and always confusing, for the researcher. The 1880 manuscript census listed each farmer and the members of his household by sex and age. It also included the amount of wages paid and the number of weeks labor was hired by race. There is no question, therefore, when it comes to estimating the number of hired hands employed by croppers and fixed renters, since any farmer paying wages in excess of sixteen weeks out of the year may be regarded as having hired a hand or hands to help him make the crop. By estimating the going wage and dividing the amount into the total wages paid, it is possible to determine the approximate number of hands so employed. But then the problem begins: what about those hands listed in the household as living with the group but not as family members? Were these individuals hired hands? Possibly, but they may have been hands hired for that year, the summer of 1880, and not laborers contributing to the crop in 1879. In addition, the enumerators commonly left blank the wages column of the owner-operators, thus making it nearly impossible to assess that group's relative efficiency.

In light of these facts, the aggregating process must be approached with caution. Accordingly, this study has attempted to gain some insight into the matter of the freedman's relative, personal efficiency by assuming (1) that the people listed in the farmer's household were either hired hands or family members who worked full time on the crops; (2) that additional hands (not in the household listing) were hired if the census enumerator listed in the "weeks hired by race column" numbers in excess of sixteen weeks; and (3) that both sexes labored with equal results in the patches. Although the results of such a procedure (especially 1 and 2) may bias the labor input data upward, it is felt that the above assumptions will not seriously skew our conclusions. First, the fact that the census itself made a distinction between those in the household and the amount of money paid for hired laborers suggests that anyone living in the one- or two-room cabins of the freedmen was more than a casual laborer or a seasonal hand. These nonfamily members of the household were most likely members of an extended family of cousins and suitors and associates who were like family but not enough so to suit the hard distinctions set up in the census. Second, unless the results of our investigation suggest that the freedmen were personally less efficient when compared to black slaves in cotton production, there is little immediate reason to be much concerned with the upward bias. If this approach suggests that the freedmen were personally more efficient or at least equaled the personal efficiency of slaves, then the upward bias would only strengthen that suggestion.[12]

In calculating investments per worker, it is crucial to limit our measures to those farmers listed as owning tools and animals so as not to compute the average investment per worker on the basis of the district's entire working population. If those farmers who appeared in the census as owning neither tools nor animals are included in the sums of labor inputs, the figures on animals and tools employed per worker would be distorted. Although this procedure substantially reduces the sample, it is assumed that, for those owning animals and tools, the number of work animals and value of tools listed probably varied only slightly from the actual average animals and tools employed per worker. It is difficult to believe that many freedmen owned a substantial surplus. When we look at the figures on animals, for instance, analysis of the manuscript census indicates that of those croppers and renters owning horses and oxen and mules—33 percent of the total listed—the average distribution only slightly exceeded the distribution reached in slavery. (The distribution had gone from 0.7 to 1 in 1859 to 0.9 to 1 in 1879). Unfortunately, in regard to tools, there is no reliable method for separating the value of plantation equipment from the value of tools listed in 1879. About the most that can be done in this matter is to note that the average value of tools per worker ranged from $6.86 for sharecroppers to $1.20 for fixed rent tenants in 1879, and to suggest that these amounts were undoubtedly below or only slightly above the per worker value of tools employed by slaves in 1859.

Finally, the nature of the problem demands finding some method for working with the total crops produced in the two periods so as to measure the per worker productivity levels on a comparative basis. Corn and potatoes and cotton must be treated as one crop. The most direct method is to compute the acreage in corn and potatoes as if it had been in cotton yielding the average bales actually produced per acre on the cotton lands. Although this method may overestimate the cotton yields, since the lands planted in potatoes and corn were probably less productive than the cotton lands, the bias, being true for both periods, should cancel itself out in the comparison. There is a problem, however, in obtaining data on the number of acres planted in corn and potatoes in 1859 inasmuch as the census failed to record the amounts of land devoted to each crop separate from the total acres improved. But as the 1880 census did report the acreage devoted to each crop farmed, it is suitable to use the average returns of corn and potatoes per acre in that census as a reliable means of determining the number of acres devoted to corn and potatoes in 1859. Little changed in the technology of corn and potato production in the time span involved, and it can be assumed that farmers utilized the less productive lands for their corn and subsistence potatoes in both periods. Accordingly, the district as a whole produced 16 bushels of corn per acre and 25 bushels of potatoes (Irish and sweet) per acre in 1879. When these yields are divided into the total amounts of corn and potatoes produced in 1859, we find that the district devoted approximately 63,085 of its improved acres to corn and potatoes in the year before the war. If it is also assumed that these acres would have returned at least the average number of bales of cotton produced per acre in 1879—0.83 bales per acre (a reasonable assumption since 0.73 bales per acre is the amount obtained after subtracting the acres in corn and potatoes from the total improved lands shown in the 1860 census)—we are dealing with about 52,360 bales of cotton, or a total of 146,119 bales, when the actual cotton yields are included in the figure.[13]

THE RELATIVE EFFICIENCY OF LABOR

With these thoughts on procedure in mind, we may go to the problem. In the first place, a glance at the published census of 1880 immediately reveals that slavery as a system of farming outproduced sharecropping in cotton and corn by a large margin. Specifically, slavery produced 93,759 bales of cotton (400-pound bales) and 904,850 bushels of corn in comparison to the 61,911 bales (400-pound bales) and 237,759 bushels produced in 1879. When corn is adjusted to its equivalent cotton yields and measured along with cotton in relation to the district's black population in the two periods, it is clear that the per capita performance of the postbellum system fell significantly below that of slavery. Slavery returned approximately 6.2 bales of cotton per slave in comparison to the 3.8 bales produced per freedman in 1879.[14]

On the other hand, a closer study of the records suggests that the 1880 system of land tenure produced more cotton and corn per improved acre than had slavery. When calculated on the basis of acres improved and with corn and potatoes adjusted to their cotton equivalents, it is observed that the postbellum system farmed less than half as many improved acres (91,013 compared to 190,800) and yet produced more than half the antebellum total (75,347 bales compared to 146,119 bales). More precisely, the district as a whole produced 8/10 of a bale per improved acre in 1879 in contrast to slavery's 7/10 of a bale in 1859. Since farmers in the postbellum era farmed fewer acres in comparison to 1859, it is undoubtedly folly to evaluate the freedman's performance on a per capita basis, as if the entire black population were fully employed in 1879. In fact, judging from the manuscript census, large segments of the black population were either unemployed or only marginally employed as pickers in 1879. What this suggests, of course, is that, while slavery had more fully utilized the district's land and labor resources as compared with sharecropping, the individual freedman equaled or possibly exceeded the personal efficiency levels achieved by workers in slavery.

But when the data are analyzed on the basis of the manuscript returns where it is possible to specify actual labor inputs per farm, it is seen that

Good and Faithful Labor

sharecroppers and fixed renters fell seriously below the per worker (ages ten to sixty-five) production levels achieved in cotton and corn by slaves in the district. Table 12 also indicates that, while the freedmen's productivity had declined, so too had the input side of their production schedules. Although tool expenditures had fallen off to 25 percent of the district's antebellum total, it cannot be said that the decline in output stemmed from reduced investments, since real investments were often hidden from view if listed among the properties of the owner-operators.[15]

Table 12
Input-Output Data Per Freedman and Slave (Ages 10 to 65), Natchez District, 1859 and 1879[a]

Measures of Input-Output	Sharecroppers		Tenants		Slavery	
	A	C	A	C	A	C
Number of farms	849	744	162	80	138	157
Acres per hand	6.6	6.1	4.6	6.2	11.9	6.8
Tools per hand	$ 3.89	$ 3.04	$ 1.40	$ 5.14	$ 10.00	$ 99.41
Animal per hand	0.56	0.42	0.51	0.74	1.04	0.87
Corn per hand	18.6	1.6	13.7	21.9	48.7	57.3
Cotton per hand	3.2	6.9	2.6	3.9	3.7	7.3

	Concordia's Ward One in 1879	Plantations Over Fifty Slaves in 1859	
		Adams	Concordia
Number of farms	229	67	88
Acres per hand	9.9	9.5	6.4
Tools per hand	$ 37.20	$ 8.59	$ 98.92
Animals per hand	1.04	0.70	0.74
Corn per hand	80.9	38.8	56.6
Cotton per hand	8.1	3.1	7.6

[a]Production is in bushels for corn and in bales of 400 pounds each for cotton. The inputs are acres improved, tools and implements, and work animals limited to horses, mules, and oxen.

SOURCE: U.S. Census (1860), Manuscript Population, Slave, and Agricultural Schedules, Adams County, Mississippi, and Concordia Parish, Louisiana; (1880), Manuscript Population and Agricultural Schedules, Adams County, Mississippi, and Concordia Parish, Louisiana.

We must, therefore, probe deeper. Fortunately, the census was enumerated on a ward basis, which provides a convenient means of singling out parts of the neighborhood for closer scrutiny. Looking at the sharecroppers in Ward Two of Adams County, for instance, we see that they exceeded the average amounts of cotton produced per worker in the county in 1859 by over a bale (4.8 compared to 3.7) while producing substantial amounts of corn (31.9 bushels compared to 42.0 bushels). They utilized approximately the same number of animals per worker (0.83 to 1) as had slavery in the county in 1859 (0.87 to 1). Here, then, was a part of the neighborhood where the croppers fared quite well as a group in comparison to the performance of slaves in 1859.

More significant yet was the performance of freedmen in Concordia's first ward. Here the census enumerator's notes have enabled the historian not only to identify the plantations therein as total units, including all the data on labor and animals and tools, but also to pinpoint thereby the personal efficiency of the freedmen involved.[16] On a comparative basis, there is absolutely no question that the freedmen in Concordia's first ward equaled the average cotton production per working hand achieved in the parish under slavery. The croppers returned 8.1 bales per worker in comparison to slavery's 7.6 bales. In regard to inputs, although the croppers here invested far fewer dollars in tools and equipment per worker ($37.20) in comparison to the amounts invested in Concordia in 1859 ($99.41 per worker), the sum far exceeded the average of $3.46 invested by all other croppers in the area. Since the above sum included the values of tools listed for the owner-operators (who most likely supplied their croppers with tools), the total value of tools employed in the ward surely approximated the average amount invested in tools per worker in the district. If judged on the basis of the census, the area's croppers utilized a greater number of animals per worker than was the case in the parish in 1859.

CONCLUSION

What does all of this tell us about the freedman's personal, relative efficiency? In the first place, if the performance of croppers in Concordia's fast ward in 1879 is at all illustrative, sharecropping was not in all cases a system of land tenure which undermined the freedman's personal efficiency when compared to slavery. Here the system utilized the land

and labor as well as, if not better than, was the case in slavery. The amount of capital for tools, animals, and land compared favorably, and the resulting production of cotton and corn actually exceeded the average levels reached on the most productive plantations in the parish in 1859. The reason, of course, is that Ward One, in the words of the census enumerator, was "considered to be the best improved and most fertile portion of the Parish." The best delta lands and rich meadows attracted significant amounts of capital regardless of the system of land tenure and organization of labor. With capital, land, and animals available in amounts comparable to the case in 1859, the freedmen's personal efficiency—defined here as a measure of how well they worked with what they had to work with—unquestionably equaled the slaves' personal efficiency. They may not have improved much under sharecropping, but it is hardly justifiable to say that the croppers were any less efficient.

In the second place, the experience here suggests that the allocation of land per worker played a crucial part in determining the freedman's relative productivity. The fact that the ward's freedmen worked 9.9 acres per worker, on the one hand, stands out in stark contrast to the district's average employment of 5.9 acres per worker in 1879 and, on the other hand, compares favorably to the 9.3 acres per worker in 1859.[17] What this suggests, of course, is that the freedman's relative productivity was largely determined by the number of acres employed. According to our sampling from the manuscript census for 1879, freedmen sharecroppers and tenants worked an average of 5.9 improved acres, and it is not difficult to suggest that had they worked the antebellum average of 9.3 acres, their product might have outdistanced the slave's productivity.[18] The calculation is simple enough: 5.9 improved acres produced approximately 4.8 bales of cotton per worker when corn and potatoes are adjusted to their cotton equivalents, whereas 9.3 acres would have yielded approximately 7.5 bales per worker. Compare this to the antebellum average of 7.0 bales per worker (with corn and potatoes adjusted), and there is a scant basis to suggest that the personal efficiency of black workers had deteriorated with slavery's end.[19]

With regard to whether this evidence is sufficient to conclusively lay to rest the contention that the freedman's personal efficiency had deteriorated with slavery's end, the answer is both yes and no. Yes, because once it is recognized that with the labor inputs calculated from the 1880

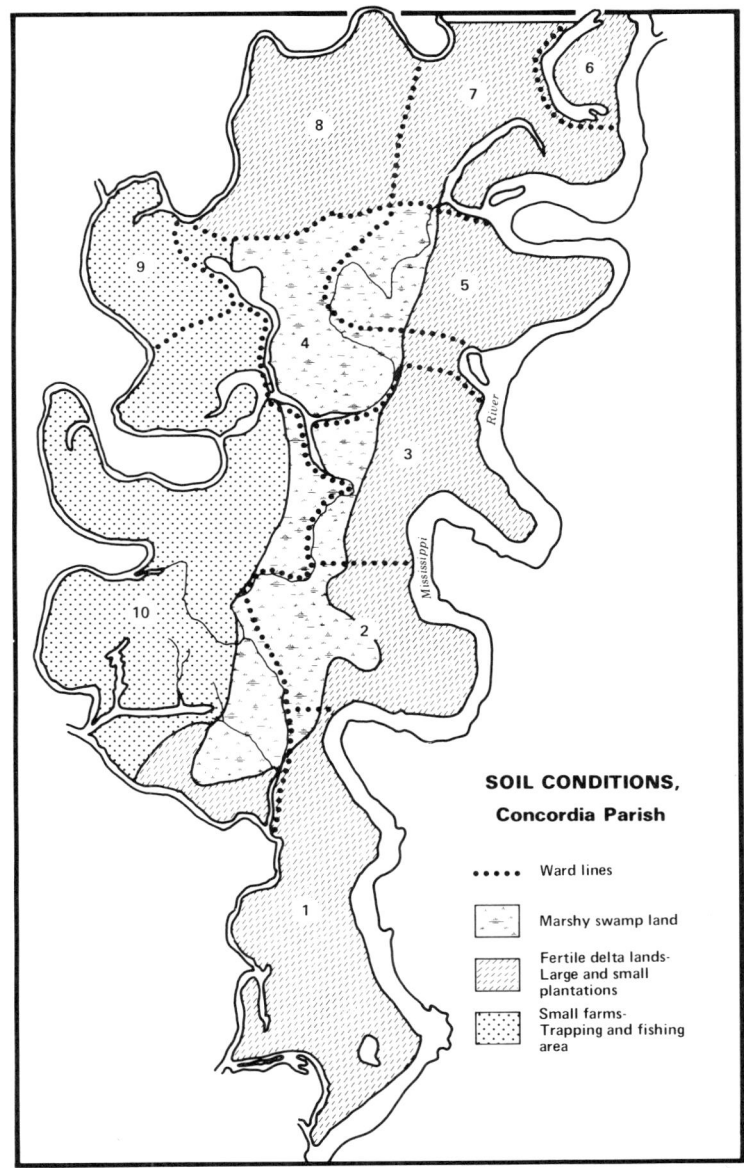

SOIL CONDITIONS,

Concordia Parish

••••• Ward lines

Marshy swamp land

Fertile delta lands-
Large and small
plantations

Small farms-
Trapping and fishing
area

Map 4. Soil Conditions, Concordia Parish, 1900.

census biased upwards (for reasons discussed earlier), we need not seriously consider the fact that the yields for 1859 are biased down-wards in our test as a result of possibly excluding some of the actual farm production. The upward labor bias of our procedure strengthens the suggestion that the freedman's personal efficiency deteriorated hardly at all under sharecropping in a comparative sense. Had fewer hands been employed, the comparison would be even more decisive in favor of the freedmen. On the other hand, the above results fail to dispose of the theory that fewer blacks and acres were employed in 1879 in comparison to 1859 because large parts of the population were simply unable to work efficiently as free men. According to this premise, those blacks actually farming were the best of the lot working the best of the lands. In this sense, slavery's end so undermined the freedmen's per-sonal efficiency as to render large numbers of them unemployable.

There is little evidence for this view, however. The system of land tenure in use in 1880 certainly employed fewer acres and people than had its antebellum predecessor, but not because of the freedman's inef-ficiency. Rather, the nature of the district's economy provided planters with fewer incentives to employ the district's entire black population in cotton production, and it gave freedmen the opportunity to pursue what economists call "nonpecuniary" goals. It was not that the price of cotton had declined to the point where planting was no longer profitable, but that slavery's end also ended the need to keep the entire black popula-tion at work. The matter was simple enough. Blacks in 1860 were not only laborers but also sources of credit and valuable commodities. Hence, a planter was motivated to own as many slaves as possible as long as the returns from planting covered production expenses, interest on loans, and costs of living. But with blacks no longer commodities in 1880, planters had less incentive to work the marginal lands or to expand their operations in order to cover the costs of owning slaves. In this context, landlords and planters may indeed have contracted with only the best (most faithful) hands and worked only the best soils in 1880 (1) in recognition that a large surplus of workers existed for hire as seasonal pickers and (2) because the expenses of cultivating marginal lands were no longer offset by the income (either cash or credit) derived from owning slaves.[20]

But in one sense sharecropping may have rendered some freedmen unemployable. The data suggest that perhaps one-fourth of the adult black population in the district lived and worked as casual laborers in the cotton economy. It seems that hundreds of people worked only as pickers or as hands occasionally brought in for cash wages to supplement a cropper's labor force in emergencies. Undoubtedly, some of these marginal hands were undesirable as full-time hands for one reason or another. Perhaps they were lame or known troublemakers. But the fact that planter-landlords were always short of labor suggests that some, possibly many, of these people preferred such seasonal work over the life of a cropper who ended his year in debt to the store. Cash wages three months out of the year, supplemented by a little fishing, a little woodcutting, odd jobs, and hunting may have provided incomes for some that equaled the incomes of croppers. Although the choice threatened its takers with insecurity and was thus unavailable to the majority of black families in the district, the fact that sharecropping had reduced hundreds of freedmen to paupers drove others to avoid the trap if at all possible. The low returns for their labor, rather than the landlord's refusal to hire them, and the fact that they were free men, combined to create a class of unemployed blacks who hardly fitted the category of "good and faithful labor."

NOTES

1. U.S. Census (1860), Manuscript Population, Slave, and Agricultural Schedules, Concordia Parish, Louisiana.

2. U.S. Census (1880), Manuscript Population and Agricultural Schedules. In this chapter, the terms used in assessing productivity and efficiency (worker and hand) refer to males and females ages ten to sixty-five years of age.

3. Ibid. (1860).

4. Ibid. (1880).

5. Ibid.

6. Ibid.

7. Ibid., Adams County, Mississippi.

8. Ibid.

9. See the introductory chapter of this study for background on the argument that the black's personal, relative efficiency had suffered with the ending of slavery and the emergence of sharecropping

10. The job of transcribing data from the manuscript census involved several steps of organization and a tedious procedure. The various census schedules, handwritten, were usually out of order in regard to the listing of individuals in

the population and agricultural sheets. That is, the individual's name would appear among a certain group, perhaps on page two, in the agricultural data and among another group on a different page in the population sheets. The task, therefore, of matching names alone was tiresome but usually possible if done with great care and diligence. Once the data were gathered for all the individuals concerned, the equally painstaking job of its manipulation was especially complex and involved.

11. More specifically, landlords and merchant suppliers were seldom listed along with their tenants or cropper customers and laborers. Thus, the monies listed as the value of tools owned by landlords could seldom be linked with certainty to the individually listed freedmen. In some instances, the tools listed were duplicated in the census as landlords claimed the implements as well as the freedmen. The problem reflected the frequency with which planters and merchants, although never abandoning legal ownership until the final installments were paid, sold tools and animals to croppers and tenants on credit.

12. The major methodological problem here is the assumed labor participation rate. The method herein employed views prime hands, ages ten to sixty-five, as the labor input units. Obviously, slaveowners might have worked women and children much longer and harder than they worked in freedom. In addition, it is doubtful that all prime hands on the plantation grew crops. Nor is it safe to assume that all family members worked full time in the fields after slavery's end. Ransom and Sutch estimate that the absolute reduction in annual labor time was in the range of five hundred to six hundred hours for freedmen after the war. Charles Seagrave notes declines of between nine and seventy-four days worked per year by Louisiana class 1 field hands in the immediate postwar years. David and Temin generally agree with these estimates, except to imply that the time difference may have been less by 1880. It is, therefore, the weight of the opinion that the methodology assumed herein overestimates the labor input of free blacks and underestimates that of slaves. The only justification for proceeding on this basis is in the logic of the problem. If the freedman's relative efficiency is high given the overestimations involved, then the insight afforded is substantial indeed. Moreover, an unwritten part of the conceptual framework underlying much of this analysis is the assumption that the actual work requirements in cotton production, given the technology of the time, were relatively set, with the crop yields unresponsive to additional labor time inputs alone. See Paul A. David and Peter Temin, "Slavery: The Progressive Institution?," in Paul A. David, Herbert G. Gutman, Richard Sutch, Peter Temin, and Gavin Wright, eds., *Reckoning with Slavery: A Critical Study in the Quantitative History of American Negro Slavery* (New York: Oxford University Press, 1976); Ransom and Sutch, "The Impact of the Civil War and of Emancipation on Southern Agriculture," *Exploration in Economic History* 12 (January 1975), pp.

1-28; Charles E. Seagrave, "The Southern Negro Agricultural Worker: 1850-1870" (Ph.D. dissertation, Stanford University, 1971).

13. The above figures are based on the district's total production and acres listed in the published census, rather than on a sample of the manuscript data. See U.S. Bureau of the Census, *Eighth Census of the United States: 1860. Agriculture*, 3: 255; *Population*, 1: 393, 397.

14. *U.S. Census, 1860*, 3: 255; 1: 118; *U.S. Census, 1880*, 3: 227,231; 1: 393, 397.

15. This point relates to what was said earlier in Note 11 of this chapter.

16. Concordia's first ward is one of the few wards in the district where the enumerator, W. F. Lanning, listed the sharecroppers by plantation and also identified the landlord of each. It is thereby possible to determine the exact amount of labor (number of hands), tools, animals, and acres employed, and crops produced on each place, subject to the assumptions mentioned earlier regarding hired hands. Moreover, all of the freedmen listed in this ward were sharecroppers.

17. These figures are based on a large sampling of the manuscript census data for the two periods. In antebellum materials, the sample equaled 92 percent of the farms listed in the district. Although the sample for 1879 was a much smaller one because of the relatively poor quality of the data—52 percent of the total cotton produced—the average production of cotton per acre derived from the sample exactly equaled the average for the total district (0.8 bales to 1).

18. With corn and potatoes adjusted to their cotton equivalents, the district returned 0.82 bales of cotton (400-pound bales) per acre in 1879 and 0.76 bales per acre in 1859. The basis for the adjustments is explained earlier in this chapter.

19. This particular argument assumes that the additional units of land per worker in the move from 5.9 acres to 9.3 could be worked with no reduction of production and labor on the previous units of land. Rather, the freedman, if the data on time lost in freedom are correct, as suggested in Note 12, probably worked the land intensively only in the sense of working fewer acres than he had as a slave. His actual labor inputs per acre, the intensity of his work, only appear to later historians to have increased because of the freedman's inability to obtain more lands to farm. Allowed a limited acreage, possibly as a means of social control, freedmen obtained maximum yields per acre and thus farmed the land intensively from the economist's point of view. Yet, additional lands in cotton, given the actual low labor input required, would not have resulted in a reduction of the yields previously achieved. It is assumed herein that individual freedmen could have worked additional acres with a proportionate increase in output up to the amounts worked in slavery. Any diminishing returns would occur at or beyond the number of acres worked in slavery.

20. Another way of reaching the same conclusion as argued here is to note the fact that after the war, planters were interested in farming their lands intensively, whereas before the war, slave masters wanted to maximize output per slave.

• 7 •

THE WORLD OF SHARECROPPING

Former slave masters in the postwar South commonly believed that black people would perish in freedom, at least as a stable and productive group of people. A correspondent to *The Southern Cultivator*, in April 1869, noted the few children at work on his plantations and predicted that "in 25 years, a negro will be almost as great a curiosity as an Indian."[1] Planter James E. Old, of Lake Providence, Louisiana, agreed, pinpointing the cause in the "miscarriages and abortions" that had "become so popular and dear to the Negro women as freedom itself." Unless these things change," he concluded, "the Negro race will, in marvellous short time, become extinct in the South."[2] The editor of the *Natchez Democrat* told his readers in 1866 that "no supernatural vision" was needed to see that the "child is already born who will behold the last negro in the State of Mississippi."[3]

For most planters, the conviction that the black race faced extinction justified their role as slave masters in antebellum times and reinforced their largely unaltered view of the black person's inability to live a normal family life in freedom.[4] "With no one to provide for the aged and young," explained the above editor, "the sick and the helpless incompetent to provide for themselves, and brought unprepared into competition with the superior intelligence, tact, and muscle of free white labor, they [the ex-slaves] will surely and speedily perish."[5] From this perspective, the conditions of freedom would so undermine the fragile social bondage holding blacks together that the very race was doomed to disappear, though the old and the young would be the first to suffer. The free black family would fail both as a means of providing for its dependents and as a means of training disciplined workers able to earn their bread by the sweat of their brows. Accordingly, the slave's family structure was viewed to have been an enforced one dependent upon the master's role and not natural to black people. After slavery, all evidence of blacks wanting to strengthen their family bonds through legalized marriage, by means of black women becoming homemakers, or in the reuniting of separated family members, was interpreted as just so much aping of white civilization. In the end, the innate weakness of the black's family structure, once it was removed from the stable community of the slave plantation, would be their undoing.

Much of the opinion on this matter reflected the strong desire to see freedom fail for blacks, causing many southern whites to interpret what were clear indications of family stability as evidence of disorganization and chaos. The manager of several Natchez District plantations, William B. Shields, whom we discussed earlier, was typical in his perceptions of the black family's adjustment to freedom. In 1863, Shields warned that one of his best slaves, "Cesar," planned to carry off his family to New Orleans. A month later, he noted that, although "Cesar has not gone yet," he expected other families to leave at the first opportunity. Another sixteen months passed with the now "perfectly faithful" Cesar still at work. In December 1865, Shields worried about the freedmen's demands for schools to educate their children. In 1866, he expressed concern about the many freedmen who were giving notice of their intentions to leave for "their former homes" in Maryland and Virginia, and about his inability to shift and relocate the freedmen among the plantations. He warned his employer that "they will all leave" should they be forced to relocate, and he noted that they remained "partly to be at home and principally because they don't trust our neighbors."[6] But the old manager was especially perplexed by the family role adopted by black women. He accused them of being generally lazy and using the pretext of their pregnancies to stop work whenever they could.[7]

Yet, Shields never understood his own observations. When he commented directly on the nature of the black family, he saw only deterioration and lack of binding family commitments among its members. "As I expected and predicted," he said in one letter, "Harriet left me a few nights after Johnson's departure. She is a very bad girl and proved very unfaithful both to her master and her husband."[8] By implication, nothing of this kind would have happened in slavery. A year into freedom, Shields wrote of the return of a former slave woman, who "came and begged" him to take her son. "She was about to abandon him." Shields decided "on the score of economy alone" to apprentice the lad and was only sorry that his mother had not come sooner as "this boy would have suited" him "better than Edgar."[9] For Shields there was nothing supportive to be said about the obvious trauma of mothers having to choose between abandoning their children or giving them up to masters from whom they had once fled; about the opportunity of mates to leave spouses bound to them in slavery for those of their own choosing; or about the freedom of women to leave the fields when pregnant. For Shields and countless others like him, the important fact was that freedom found blacks abandoning their children, wives their husbands, servants their masters, and workers their duties in what was interpreted as the withering away of not only the black family but also the mutually supportive family of the slave plantation.

Anna is not only willing to go to you but she says she is anxious to do so—but has no one here with whom to trust her two children. Times have changed, fathers, mothers, husbands, wives, brothers, and sisters do not care to assist each other or their own children much less the children of others. She says she will go at any moment you say so, but must be allowed to take her children with her. To leave them here would be, under the new regime, equivalent to abandonment. So she says.[10]

It is not that Shields completely misread what he saw. Historian Herbert Gutman has written that the slave community's viability rested partly upon its embrace of relatives and friends as common offspring, offering them the shelter and compassion of genuine parental affections.[11] Perhaps this is what Shields sensed was different in freedom. With the breakup of the slave plantation, people who had been members of a community of "fellow servants" were less tightly knit and relatively free to concentrate their efforts on themselves and on their immediate families. In some cases, those elderly and young with no immediate families in slavery were left behind when slaves fled the plantations. Slaves with families on different plantations might have been the first to leave in hopes of finding their loved ones. In the Natchez District, from the moment U.S. troops arrived, a series of related conditions undermined the chance for much family stability among blacks. As noted earlier in this study, during the war the U.S. Army housed blacks in refugee camps, directed others to labor on abandoned and leased plantations, and drafted able-bodied males into black military units as a kind of home guard. All of these measures undermined existing family ties and proved frustrating to refugees eager to be reunited with their desperate family members.

The refugee camps in time became displacement centers where the sick, young, and elderly were located, usually apart from former friends and family. While these camps provided some shelter and minimum care for people unable to help themselves, they also separated blacks from their families, thereby adding to their sense of displacement. It is likely that at the urging of the camp officials some parents left their families or some of their people in order to work on nearby plantations, and they intended to return for them once they were settled or as soon as possible. Whatever the reason may have been, the high percentage of children in the camps suggests that the young were indeed among the displaced, although the fact that 45 to 50 percent of a camp's population were adults makes it difficult to argue that the camp children had been abandoned.[12]

More disruptive yet was the army's policy of enlisting able-bodied black males as soldiers and of placing their women, children, and elderly kin at work on the leased plantations.[13] Unfortunately, the army's inability to protect the plantations and the primary importance of having a sufficient and stable labor force at work resulted in disaster. Women and children were frequently taken back into slavery or killed by rebel forces preying on the Concordia plantations. And if that were not enough, black soldiers experienced more than a little difficulty in rejoining their families at the end of their service. Some planters refused to contract with the former soldiers, and black women had difficulty breaking their contracts to join returning husbands and fathers.[14] This situation reached its extreme in 1866, during the brief period of the "Black Codes." If for any reason during that time blacks found themselves out of work, either because they had been discharged, were between contracts, were waiting for their husbands and fathers, or they had refused to sign contracts under the terms offered, they were subject to arrest as vagrants and a period of service for any planter paying their fines or giving bond.[15] The "Black Codes" may not have broken up many existing families, mainly because they were seldom enforced in the district, but they undoubtedly prevented the reuniting of families broken up in slavery. Especially for those young adults and children separated from their families either by slavery or by the circumstances of war, the "Black Codes" placed them at the mercy of every white citizen in the district. If the local sheriff reported a minor as a vagrant, the young person found himself subject to the whims of a local probate-judge. In the Natchez District, this possibility spread panic among the neighborhood blacks. The Freedmen's Bureau reported a steady registry of complaints and numerous petitions by the freedmen for their children, with many "offering to give bonds for their children's support and care."[16]

To add to the confusion, in the postwar years the Natchez District was engulfed by freedmen brought into the area as laborers. Employment agencies in New Orleans did a healthy business bringing young men and women to work on the Mississippi and Louisiana plantations. Others were brought in by the U.S. Army from Alabama, Tennessee, and upland Mississippi and Georgia to pick cotton in the swamps. Although we have no way of estimating the total number brought into the district in this way, it was substantial, and most of the workers resembled a migrant labor force of individuals with families and friends elsewhere. The end result was an unsettling effect upon the local black inhabitants, and more than a few uprooted members of district families were lost in the shuffle.[17]

Important too was the continued terrorization of blacks as a social policy aimed at keeping them from exercising free movement. Neighboring areas to Natchez, those somewhat distant from a well-equipped Freedmen's Bureau headquarters, were commonly patrolled by a "black cavalry" of white militia. In several instances, these bands "brought back" freedmen who had gone to another county, possibly the Natchez District, "whipped them and ordered them not to leave again."[18] Although such activity occurred less frequently in Adams County, it was so widespread throughout the adjoining areas, including across the river in the delta, that it probably kept more than a few district freedmen from being reunited with family members scattered during the war years.[19]

Of more certain impact on the stability of the black family was the handing over of convicted black criminals to planters who had paid their fines or posted bonds. This practice made no distinction in regard to the criminal's family status. Unlike the unemployed of the "Black Codes," these people were not vagrants but convicted felons, usually sentenced for the crime of stealing cotton. They were arrested in the late fall, convicted in midwinter, and sentenced at the start of the new contracting season. The punishment was almost always a fine of between $10 and $50. If the black lawbreaker was unable to pay the fine, a jail term of two to three months was applied. On a single day in February 1866, the Adams County court convicted fifteen freedmen for such petty larceny. "A number of those convicted and sentenced have been bought out by parties in want of laborers, and the remainder will doubtless be disposed of in like manner. Those wishing laborers and willing to dispense with references as to character, need only call at the Sheriff's office, pay the fine and expenses of the above and take them home."[20] The fate of these freedmen was most uncertain, given the likelihood that they ended the year in debt to the planter's store for rations and supplies, and since, as convicted criminals, they were outside the small protective authority of the Freedmen's Bureau. The courts handled dozens of allegations and convictions for cotton stealing, and this was the beginning of a kind of enforced peonage that was soon institutionalized in the convict and bound labor penal system for which the South became notorious.

THE BLACK FAMILY IN THE POSTBELLUM ERA

In spite of all these obstacles, the black family emerged from the war a force to be reckoned with. In the first few years of their freedom, blacks contested all attempts to reintroduce the centralized plantation system of gang labor, persistently called for schools to educate their children, rushed to have their marriages legally recognized, and worked to set themselves up in as near a family farming situation as was possible, given their poverty and lack of government support. Within the first generation after slavery, the freedman's commitment to the family as a way of life and labor had been largely realized. By 1880, the typical black agricultural worker in the district lived as a member of a nuclear family. In Concordia Parish, 62 percent of all agricultural blacks were either parents or children in a family limited to immediate family members. Another 23 percent lived in families with uncles and aunts and elderly kinfolk. Only 15 percent lived in what could not be described as a family unit.[21] To be sure, such a family structure indicated the extent to which the system of sharecropping had fashioned itself to accommodate the close kinship ties of black people. In their classic study, *The Deep South*, Addison Davis, Burleigh B. Gardner, and Mary Gardner describe the role of the black family structure in sharecropping. According to their findings, nearly 40 percent of all agricultural workers in their "Rural County" were the women and children of the sharecropper or tenant male under contract. Then, as in the 1880s, the picking season and the "making of the crop" required all available hands in the fields. Landlords were thus unwilling to rent to unmarried men and gave preference to tenants who had a young or middle-aged wife with children living in the family.[22] This practice had started early and by 1880 was a central feature of the system of agricultural labor in the district. It may explain the small number of people in the work force who were not living with or were not themselves members of a family. Even before sharecropping had been fully institutionalized in the district, freedmen often contracted in squads with their families as the basic members of the labor unit. But it cannot be assumed that the eventual use of the family as the main component of labor on each tenant patch reflected the landlord's unwillingness to rent to any but black families. Rather, as discussed earlier in this study, district blacks would have it no other way, and landlords eventually adopted the system, even preferring it to a gang labor arrangement.

How did the black family compare in freedom to the black family in slavery? The fact that freedmen insisted on the family structure in their contracting tells us little about what they had experienced in slavery. Fortunately, in 1859 the census enumerators for Concordia Parish gathered their information cabin by cabin, recording the age and sex of the slave inhabitants. In 1860, 36 percent of Concordia's slaves lived in cabins where the age and sex composition of the residents indicates that they were family groups of parents and children (two adults of the opposite sex, compatible in age, and one or more children, compatible in age). Another 24 percent lived in cabins of the family type described above, with the exception that more than two adults resided therein. These were most likely kinship families of aunts, uncles, and cousins. Still another 10 percent lived in cabins of one adult and children, or so-called one-parent households, and 3 percent lived in family cabins with elderly people. If these were families, the comparison with 1880 is very revealing. A much larger percentage of blacks in freedom lived in

Plate 10. A Freedman's Home.

families limited to parents and children than was the case in slavery. This meant that many more blacks in slavery lived in extended families than was the case in 1880. A relatively small percentage of blacks lived alone in either year, and about the same ratio lived in one-parent situations in 1880 as in 1860. On the other hand, a significantly larger percentage of blacks lived in heterosexual arrangements in freedom, while a substantial number of slaves lived in totally adult quarters in comparison to the living practices of the free agricultural blacks in 1880.[23]

What blacks had gained in freedom, at least according to the manuscript data for Concordia Parish, was a significantly strengthened nuclear family structure (Table 13). It would be too much to assume that all slaves who had lived in cabins with parents and children were related, but most probably were, and others may in time have become so-called fictive cousins and aunts. With freedom, the nuclear family

Table 13
The Black Family in 1860 and 1880

Mother, father, children	36%	62%
Mother, father, children plus other adults	24	12
One adult and children	10	8
Family and elderly	3	3
Black Family (Total)	73	85
Male and female	3	7
Two adults or more	18	5
One adult	1	2
Elderly	1	0
Other	4	1
Total	27	15

SOURCE: Manuscript Census.

structure emerged dominant. And if in their rush for freedom in 1863-1865, parents and adults had abandoned children and the elderly, it was not apparent by 1880. As in slavery, most of the elderly lived in family situations, and even fewer lived alone in the 1880 period. Plantation slavery, it is important to recall, rested upon what appears to have been a permanent family structure, although it may have nourished a plantation community of kin and kin-like friends as much as it nourished the nuclear family unit. But the family unit was the basic living experience of Concordia slaves. When freedom scattered the plantation slaves into their relatively isolated family units, it separated them in space for the first time as the old quarters were broken up into the separate family cabins of sharecropping. But the distribution of blacks among their dwellings changed little over time, with the average cabin holding about five people in both periods. The difference, of course, was that the freedmen were much more removed from the watchful eye of the plantation owner, even as they were more removed from their neighbors and kin. Thus, the black community, which had rested upon both the nuclear and extended family cabin in slavery, had become less localized and less community rooted in freedom. Such a shift was of profound consequence for the people living through it.

A NEW DEPENDENCY

For one thing, the isolated family experience of sharecropping may have contributed to a new kind of dependency for district blacks. Although some form of black community survived after slavery as freedmen shared the common experience of their close slave associations, of being black and poor, and of being related in blood and through marriage, and as they still lived in the neighborhood of their antebellum slave plantations, often in sight of one another, they were now more apart from their kin and fellows than ever before. Most black families stayed in the district, but few stayed on the same farm for enough time so as to establish the rootedness to a specific place that would have strengthened the extended family ties of their slave community. Some freedmen managed to farm in the same general area or on the same plantation for years at a time, but they were few in number. Shepard Bershey, for instance, had contracted after slavery with his old master as foreman of a squad of six freedmen. In a few years he moved to sharecropping for the same landlord and was still on the place in 1885 farming as a fixed rent tenant for his former master's heirs. In all that time, Bershey's material condition had changed but little. In 1868, he ended the year owing his landlord $11.28. He was still in debt in 1885 when he farmed "the same portion of Hollywood as last year" as a renter, paying 950 pounds of lint cotton and seed. The only thing atypical about Bershey is that he had stayed on the same plantation for all those years.[24]

The average black family moved from place to place, from landlord to landlord, and even from supplier to supplier without ever leaving the neighborhood. Merchant Isaac Friedler's accounts in 1880 show only two of the five accounts started in 1874 on the Minorca plantation still running, one of his five on Palo Alto, two of his fifteen on Weaverly, and one of his seven on Armandellia.[25] Perhaps the missing families were still on the same plantations and were contracting with other merchants, but the lien and plantation records suggest differently. The freedmen had probably moved with their families from year to year and from place to place in a pattern of displacement that undoubtedly weakened the extended family ties of slavery.

Not surprisingly, the displacement of blacks upon the collapse of slavery found the freedman's commitment to family, strong as it was in slavery, greatly strengthened in freedom. The first generation after slavery, it is to be remembered, was a period of intense uncertainty and struggle for district blacks. The one thing they had won was the security of their families. Having nothing else, the black family became for many their only source of refuge. Yet, insofar as the black family nourished the independence of black people, it may also have partially supported the growing caste definition of the black's world. By the twentieth century, district blacks had become a subordinate caste of people hopelessly subjugated to all whites as an inferior people. In the transition period between slavery and sharecropping, the contours of such subjugation were eminently visible, but the process was still somewhat dynamic. It may have been that the freedman's struggle to secure the integrity and security of his family, his struggle to separate himself and his loved ones from the daily control of the white master, left him willing to compromise on the less immediate demands of caste subordination—a subordination of deference and reticence in the presence of whites. The very refuge of the black family enabled freedmen to tolerate a caste subordination which they were powerless to prevent. What must be understood is that no matter how poor and exploited the first generation of freedmen was, they had put behind forever the threat of being sold or of having family members torn away by the slave trade. In addition, the typical black sharecropper worked relatively free of white supervision. Except in a few cases, the humiliation of constant white overlording was also a thing of the past. This reality, the fruit of the freedman's struggle in the generation after slavery, gave all district blacks a stake in the society and economy of sharecropping that may not be overlooked.

There was another way in which the black family may have served to weaken the freedmen's resistance to their caste subordination. Eugene Genovese has argued that because the collective resistance of slaves to their paternalistic masters was "defensive" and aimed at "protecting individuals against aggression and abuse; it could not readily pass into an effective weapon for [class] liberation."[26] In the first generation after slavery, however, blacks clearly and actively resisted the dictates of their former white masters both collectively and individually. Much of this resistance took the form of entire plantation communities refusing to work under the old conditions or in the old ways. In the process of this refusal and collective resistance, the freedmen procured the security of their families in open acts of defiance, and, again in the process of their collective resistance, they separated and moved apart. Had the antebellum black community of slavery somehow continued to survive in freedom—characterized as it was by its geographic rootedness, by its provisions for mutual aid, and by its close kinship relations—a basis for an effective collective confrontation with the political economy of caste and class might have grown and spread. As it was, the shift from the extended community structure of slavery to the displaced nuclear family of sharecropping left blacks relatively isolated and thus vulnerable to the new dependency of the coming generations.

Without farms and homes of their own, hopelessly impoverished, and unable to counter the intimidation tactics of the dominant white and racist society, Natchez District blacks were soon caught up in the caste and lower class confines common to all southern blacks in the hundred years after slavery. But this new dependency was not the same as slavery. It is true that the freedmen were deeply in debt, and in that way they were tied to the landlords and merchants who controlled their lands and their food. They were also forced to practice absolute deference in the presence of all whites. Nonetheless, blacks never functioned as the good and faithful laborers their former masters remembered them to have been in slavery. They challenged their landlords in 1880 by moving, by stealing cotton, by working at their own pace, and they did the same in the 1930s. They held their labor off the market in 1868 and in

1880, and they did the same in the 1930s—not collectively but individually. They did so in protest of the planter's meanness or obvious cheating, while all the while showing deference and feigned respect. One old sharecropper in 1934 showed his link to his grandparents in saying:

A lot o' people won't stay on eny place mo'n one yeah. But I ain't nevah bin dat way. I don' b'lieve in movin' ev'ry yearh, lak a lot o' people. But if I tek a mine teh move, I tell you one thing—ain't nobody kin stop me! If I mek up my mine tuh move, I'm going! Don't keer ef it's in de rain. I'll go in de rain; don' pay it no mine. Pull up er wagon, an' put ev'ry stick in heah in it, an' go![27]

This was indeed a new kind of dependency.

THE NEW PATERNALISM

This new dependency of district blacks gave rise to a new paternalism among the district's white elites as well. The old southern planter class had changed in function and form after slavery, but the continued dependency of blacks perpetuated a sense of superiority and power that dominated even the marketplace. Within a generation of 1880, as a result of the unchanging dependency of blacks in sharecropping, Natchez Distict whites came to accept what they believed was the burden of their mastery. A white woman related an incident to Davis, Gardner, and Gardner in which her father had tried to get a Negro tenant to buy land. "The Negro, however, assured his landlord that he didn't want a 'place of his own.' He just wanted a buggy and a horse."[28] The woman was convinced that "they don't want to be independent." Other planters saw the new dependency as slavery reversed: "Speaking of slaves, the Negro is no longer the slave, but the planter is. We have to worry over the crop, over financing the tenant and everything like that, while he just looks to us to take care of him and hasn't a worry as long as he is fed."[29] As recently as 1976, a descendant of the district's nineteenth-century cotton masters told visitors to his home that the blacks living on his place would sell their eggs to no one but himself, though he had told them that they could get a better price in town.[30]

Obviously, this new paternalism initially reflected both the employer's needs in dealing with blacks as employees and the social demands of enforcing and cultivating a caste society in the making. Gone forever, perhaps, was the notion that freedmen could be trusted as members of the family or that the plantation was as much a home as it was a business. Yet, numerous blacks stayed attached enough to the place to be thought of as the landlord's "niggers." So obvious was this attachment that historian U. B. Phillips thought of this dependency as a poor way to run a business. That district landlords and planters hoped to combine business with this new paternalism was evident from even the earliest days of freedom. In 1868, a district freedman filed a complaint with the Freedmen's Bureau charging his former master for unpaid wages due. The freedman claimed that his employer had agreed to pay him and his wife $10 and $7 per month, respectively, and rations for 1867. But at the end of the year, the planter had refused to pay them their wages "on the ground that nothing had been made on the place." The couple then agreed to work the next year for "half of the net value of the crops raised." At year's end, the planter declined to pay the wife anything but the rations already consumed and gave the husband a "due bill" for what was coming to him from the crop. The freedman then took the matter to the bureau.[31]

The planter had his own story to tell. According to the planter, he had never formally contracted with the couple in 1867, but had kept them on the place out of sympathy, provisioning them with rations. Most of that first year, he added, the couple had been bedridden and unable to work. In 1868, the planter, though having considered the possibility of working the husband for half the crop, decided to employ him on terms left to be determined at the end of the season. When it came time for settlement, the planter gave the husband a note for $80, telling the bureau agent that "I didn't consider that I owed it to him."[32]

The outcome of this dispute is unknown, but it illustrates the nature of the new paternalism that soon engulfed district whites. the old planter had felt perfectly justified in his conviction that pay and rations, and even the eventual "due bill," were but aspects of his own generosity. The freedmen couple were free to live and labor on the place for rations, a roof over their heads, and something extra should the planter have a good year. Or they were free to go. This approach to the freedman's role in the economy and society of the Natchez District contained the seeds of the new paternalism which was soon to take root. Much of the old paternalism was still present in the planter's belief that he should take care of the freedmen and be the best judge of their interests. This attitude generally reflected the paternalism of slavery wherein the master had felt justified in keeping blacks in slavery because of the protection, Christianity, and guidance he gave them in return for their involuntary labor. What was new, however, was the planter's willingness to define his responsibility in terms of fair and honest treatment, and his willingness to admit that blacks could abandon the plantation if not satisfied with the terms offered. In the case discussed above, the planter made no attempt to justify his actions in terms of moral guidance or protection. It was now a question of fair pay and proper remuneration, though what was defined as fair and proper was never considered to be a question of negotiation.

By 1930, according to the work of Davis, Gardner, and Gardner, a new paternalism based upon a system of caste relationships had become fully institutionalized in the area. Blacks were expected to be deferential and courteous to all whites, quietly subordinate, and unquestioning of the landlord or merchant supplier's fairness and honesty in business. In return, whites left blacks largely unsupervised in their work and living, allowed them the integrity of their families, and committed themselves to dealing honestly with their black tenants at settlement time. But in being deferential and unquestioning, blacks had little means of assuring fair treatment or of having any say in the establishment of a criterion of fairness. For whites in 1930, just as it had been in 1868, "to deal honestly with blacks" meant to have them dependent enough on the store to keep them at work and subordinate as a class and caste of people. And if the new paternalism was not enough, the landlord was able to insure such deference and subordination by the use of various forms of social, legal, and physical intimidation.[33]

CONCLUSION

In the first generation after slavery, this new paternalism of so-called honest dealings found freedmen wanting contracts that would protect their claims on production. They also sought places of their own as much removed as possible from the planter's peculiar sense of justice, and they desired legal recourse for the arbitration of disputes arising out of this new form of planter benevolence. Unwilling to accept the planter's sense, or any white person's definition, of what constituted fair returns for their labor, freedmen moved from place to place, experimented with contract variations, and even held their labor off the market. But in time, as a result of their needs for shelter and supplies, their abandonment by their northern protectors, and their very commitments to family and place, Natchez District blacks found themselves bound by a new set of constraints, a new kind of dependency, and a new sort of paternalism as well. What they had won by their struggle, though not the justice and equality they had hoped to achieve, was a new status of family integrity and independence which not even the intimidation and terror of their caste-bound world could weaken. In sharecropping, district blacks had become an impoverished group of displaced workers, but their struggles in the transition period between slavery and sharecropping constituted a legacy that kept them and their descendants free-born people in the generations after slavery. It may not seem much of a victory today, but for a people born in slavery, it was quite an accomplishment.

NOTES

1. See H, "Letter from Mississippi," *Southern Cultivator* 25 (April 1869).

2. F. W. Loring and C. F. Atkinson, *Cotton Culture and the South Considered with Reference to Emigration* (Boston: A. Williams & Co., 1869), p. 6.

3. *Natchez Democrat*, January 8, 1866, p. 2.

4. See Leon Litwack's *Been in the Storm So Long: The Aftermath of Slavery* (New York: Alfred A. Knopf, 1979), pp. 361-63. Litwack correctly notes the slavemaster's view of freedom's effect on blacks, but fails to discuss how rooted it was in their feelings about the nature of the black family.

5. *Natchez Democrat*, January 8, 1866, p. 2.

6. William B. Shields to William Newton Mercer, December 11, 1863, January 25, 1864, September 20, 1865, December 12, 1865, November 7, 1866, November 18, 1866, December 1, 1866, January 6, 1867, William Newton Mercer Papers, Louisiana State University, Baton Rouge, Louisiana.

7. Ibid., January 6, 1867.

8. Ibid., January 25, 1864.

9. Ibid., November 28, 1866.

10. Ibid., March 27, 1867.

11. Herbert Gutman, *The Black Family in Slavery and Freedom, 1750-1925* (New York: Pantheon Books, 1976), pp. 97, 154, 223.

12. Major G. D. Reynolds, Provost Marshal, Natchez District, to the President of the Board of Police, Adams County, Mississippi, November 12, 1865, Records of the Bureau of Refugees, Freedmen, and Abandoned Lands (hereinafter cited as BRFAL), Record Group 105.

13. Jas. B. McPherson to Brigadier General Thomas E.G. Ransom, U.S. Army, Natchez. July 17, 1863, Official Records, Ser. 1, Vol. 24: 521.

14. General T. J. Wood, Assistant Commissioner, Bureau of Refugees, Freedmen, and Abandoned Lands for State of Mississippi, Vicksburg, Mississippi, to General O. O. Howard, Commissioner, Bureau of Refugees, Freedmen, and Abandoned Lands, May 7, 1866, BRFAL, Record Group 752.

15. *Natchez Democrat*, March 26, 1866, p. 2; U.S. Congress, Senate, T. J. Wood to O. O. Howard, October 31, 1866, 39th Cong., 2d ed. Sess., 1866, *Senate Exec. Doc.* 6, pp. 94-100.

16. Colonel S. Thomas, Assistant Commissioner, Bureau of Refugees, Freedmen, and Abandoned Lands for the District of Mississippi, to O. O. Howard, February 5, 1866, BRFAL, Record Group 752.

17. U.S. Congress, House, Testimony of Dr. James M. Turner, March 14, 1866, Joint Committee on Reconstruction, 39th Cong., 1st Sess., 1866, *H. Rept.* 30, Pt. 4, p. 127.

18. U.S. Congress, House, Testimony of Captain J. H. Mathews, Provost-Marshal and Sub-Commissioner, Bureau of Refugees, Freedmen and Abandoned Lands, Amite County, Mississippi, March 10, 1866, Joint Committee on Reconstruction, 39th Cong., 1st Sess., 1866, *H. Rept.* 30, Pt. 4, pp. 142-43.

19. U.S. Congress, Senate, T. J. Wood to O. O. Howard, October 31, 1866, 39th Cong., 2d ed. Sess., 1866, p. 4; March 15, 1866, p. 4.

20. *Natchez Democrat*, February 26, 1866, p. 4; March 15, 1866, p. 4.

21. See U.S. Census (1880), Manuscript Population Sehedules, Concordia Parish, Louisiana.

22. Allison Davis, Burleigh B. Gardner, and Mary R. Gardner, *Deep South: A Social Anthropological Study of Caste and Class* (Chicago: University of Chicago Press, 1941), p. 327.

23. U.S. Census (1860 and 1880), Manuscript Population Schedules, Concordia Parish, Louisiana.

24. Contracts, January 3, 1867, January 14, 1868, January 1, 1871, February 1, 1873, January 1, 1876, January 8, 1885, J. A. Gillespie Papers, Louisiana State University, Baton Rouge, Louisiana.

25. Liens and Mortgage Records, June 27, 1874, August 11, 1874, August 13, 1874, August 11, 1880, Concordia Parish, Louisiana, Office of Records, Vidalia, Louisiana.

26. Eugene D. Genovese, *Roll, Jordan, Roll: The World the Slaves Made* (New York: Pantheon Books, 1974), p. 5.

27. Quoted in Davis, Gardner, and Gardner, *Deep South*, p. 342.

28. Ibid., pp. 17-18.

29. Ibid., p. 19.

30. Personal conversation between author and a retired member of the district's white upper class, upon a visit to his antebellum mansion estate, April 15, 1976.

31. Office of the Assistant Commissioner, Bureau of Refugees, Freedmen, and Abandoned Lands, Natchez, Mississippi, September 7, 1868, BRFAL, Record Group, 752.

32. Ibid.

33. Ibid., pp. 392-404.

• 8 •

SUMMARY AND CONCLUSIONS

Misunderstandings about the origins of sharecropping seem so ingrained in the history and lore of the South that it may be futile even to attempt to set the record straight. The generally accepted version of its genesis has taken on the influence of a symbolic myth from which not even the practitioners of the new, so-called objective, economic history have escaped. The enduring nature of this misunderstanding is evident when comparing the writings of William Alexander Percy, perhaps the most literary of the South's planter sons in the early twentieth century, with the scholarship of historical economists Ralph Shlomowitz and Joseph D. Reid, Jr.[1] No more succinct statement of the accepted, traditional view exists than that penned by Percy in his famous autobiography, *Lanterns on the Levee*. For Percy it all began as a bargain. Ex-slaves, in the first years of their freedom, found themselves unable to cope and returned to the plantations asking for a chance to work once again in cotton. The planters, having nothing to offer "except good land and leadership," took back their former slaves on terms mutually beneficial to both:

I have land which you need, and you have muscle which I need; let's put what we've got in the same pot and call it ours. I'll give you all the land you can work, a house to live in, a garden plot and room to raise chickens, hogs, and cows if you can come by them, and all the wood you want to cut for fuel. I'll direct and oversee you. I'll get you a doctor when you are sick. Until the crop comes in I'll try to keep you from going hungry or naked insofar as I am able. I'll pay the taxes and I'll furnish the mules and plows and gear and whatever else is necessary to make a crop. This is what I promise to do. You will plant

and cultivate and gather the crop as I direct. This is what you will promise to do.
When the crop is picked, half of it will be mine and half of it yours. If I have
supplied you with money for food or clothing or anything else during the year, I
will charge it against your half of the crop. I shall handle the selling of the
cotton and the cotton seed because I know more than you do about their value.
But the corn you may sell or eat or use for feed as you like. If the price of cotton
is good, we shall both make something. If it is bad, neither of us will make
anything, but I shall probably lose the place and you will lose nothing because
you have nothing to lose. It's a hard contract these hard times for both of us, but
it's just and self-respecting and if we both do our part and have a little luck we
can both prosper under it.[2]

And so, according to legend, sharecropping emerged as planters contributed their lands and brains in return for the freedman's faithful toil.
As long as each side lived up "to the contractual obligations of the
system," added Percy, it was "one of the best systems ever devised to
give security and a chance for profit to the simple and the unskilled."[3]
Planters thus came to sharecropping, in Percy's view, largely because of
their traditional role in southern agriculture and in full understanding that
the freedmen were simply incapable of working on their own. Blacks
needed direction and productive lands to farm. Planters could give them
both, and did, once the freedmen agreed to accept the terms offered.

Modern scholars know that the story was more complex than this.
Such issues as the role of the Freedmen's Bureau, the attitude of freedmen
towards gang labor, and the inexperience of planters with free labor
were all part of the context in which the system of sharecropping emerged.
Yet, the most recent writings on the subject generally support the basic
outline suggested by Percy. Ralph Schlomowitz, for instance, believes
that sharecropping, or the collective share arrangement as he calls it,
"was widely adopted as it proved initially satisfactory to both planters
and freedmen."[4] He does differ with Percy as to the motivation of
planters and freedmen in originating the system. According to Schlomowitz,
planters considered sharecropping to be a good "group incentive scheme"
for the freedmen, which meant that the former master would spend less
time in costly supervision, thus saving money.[5] The freedmen, on the
other hand, liked sharecropping because it left them relatively free from
supervision.[6]

Although Percy would agree that sharecropping was a free transaction of sorts, he would have trouble accepting Shlomowitz's notion of

what was uppermost in the minds of the parties to the bargain. For him, planters acted out of their traditional sense of noblesse oblige, while the freedmen had simply come to their senses. In Percy's version, the coming of sharecropping is a tale of how a people went about creating a "humane, just, self-respecting, and cheerful method of earning a living."[7] In his version there is no need to emphasize as a "mechanism of transition" the economic advantage of shares (1) in postponing payday to the time of harvest, although Percy clearly acknowledges that the planters had little cash; (2) in reducing the risks borne by the planter by sharing them with labor; (3) as a group incentive for the freedmen by giving them an interest in the crop; or (4) in reducing the costs of supervision. For Percy, such advantages were undeniably present, but such market-oriented considerations were not the essential aspects of the bargain struck.

Joseph D. Reid, Jr., while agreeing that planters and freedmen freely settled upon sharecropping in accordance with the mutual advantages perceived, adds a few new twists in his view of the motivations involved.[8] In his opinion, freedmen returned to the plantation wanting to know how best to farm and, in recognition of their scant managerial skills, looked to the planter for assistance: "What better way to ensure that such assistance be accurate and timely than by giving the landlord-advisor a share of the crop?"[9] This is very close to Percy's view without the racism. But Shlomowitz's position is also somewhat addressed in Reid's suggestion that what the freedmen were after was "accurate and timely" direction, which by implication seems to mean that they were not interested in the constant and daily overlording of their work.[10]

For planters, on the other hand, working with free labor posed a fundamental problem. Their experience in plantation management was based upon the coercion of labor, and they thus had little understanding of how to deal with free labor in a supervisory but noncoercive capacity. Apparently recognizing their limitations, the former slave masters saw sharecropping as a means of providing freedmen with the needed direction while freeing themselves from the unwanted task of free labor supervision. In addition, because shares kept labor attentive to the crop through the harvest, planters were free to devote their energies and time to Reconstruction "politicking."[11]

Neither Percy, Reid, nor Shlomowitz have much evidence to support their interpretations of how sharecropping originated. Their work, rather, marshals bits and pieces of evidence from diverse sources in support of preconceived theories. For Percy, planters did what they were traditionally expected to do in regard to blacks and agriculture. For Reid and Shlomowitz, the free market guided the decisions made as freedmen and planters followed their economic self-interests. Both approaches agree that sharecropping was the product of a freely arrived at bargain reflecting the different interests involved.

Nor do the above approaches provide much insight for understanding the origins of sharecropping in the Natchez District. Here the key determinant was the freedman's refusal to work except in some type of sharecropping system or arrangement. Planters literally were dragged kicking and screaming into the system. Unable to force the freedmen to work for fixed wages in a gang setting, planters accepted sharecropping because they had no choice in the matter. Once it was forced upon them, the system carried with it many of the economic advantages described above, and landlords and merchants, now changed and affected by the system they would have avoided, worked to make the best of it. In all cases, however, sharecropping meant a reduced managerial role for the old planter class. And in few cases could the system's origins be correctly described as a mutually arrived at arrangement in which freedmen either came to their senses in the way implied by Percy or sought "timely" direction in the manner suggested by Reid.

It is not too difficult to understand the source of Percy's view. He was a planter's son reared in a society in which his version had become the accepted legend and the traditional explanation. The tale nicely explained the reemergence of the planter/landlord to the South's "top rail," while justifying the freedman's dependency and poverty without reference to any suggestion of exploitation. So deeply ingrained was this view of how it all began that it was commonly expressed even by those freedmen still living at the time of Percy's writing. In the Works Progress Administration (WPA) interviews with former slaves in the 1930s, numerous instances may be found in which still living freedmen verified Percy's story in their recollections. In most cases, the interviews had been conducted by a white person. In the Natchez District, Mrs. Edith Wyatt Moore, the author of romantic local history, did the interviewing. In the case of Charlie Davenport, nearly one hundred years old at the time, we see how well he understood what was expected of him, or the degree to which even former slaves had come to believe the accepted story:

q 1. Like all de fool niggers ob dat time I wuz right smartly bit by de freedom
bug for a while. Hit sounded powerful nice to be tole:
"You don't have to chop cotton no mow. You kin thow dat hoe down en
go fishin when evah de notion strike you. En you kin roam around at night
en cote (court) gals jest es late ez you please. Aint no marster gwine to say
to you:
  "'Charlie you is got to be back when de clock strikes nine.'"
I wuz fool enough to b'leeve all dat kind ob stuff. But to tell de honest to
God trufe most of us didn't know ourselves no bettah off. Freedom meant
we could leave whar we'd been bawn en bred, but hit also meant, we had
to scratch fur ourselves. Dem what left de ole plantation seemed so all
fired glad to git back dat I made up my mind to stay put.

q 2. I shore ain't nevah heard 'about any plantation bein divided up but I heard
a lot of yeller niggers spoutin off how dey wuz gwine to take ovah de white
folks land fur back wages. Dem bucks jest tuck all dare wages out in talk.
Cause I aint nevah seen no land divided yet.

q 3. I dont know nothing 'bout no compulsion. Guess I didn't rightly hear what
you axed me. Oh, dat hit! Did any body make me work? No, 'deed no. But
dem Yankees shore made my daddy work. Dey put a pick in his hand
instid of a gun en made him dig a big ditch in front of Vicksburg. He
worked a heap harder fur his Uncle Sam den he worked for Old Marse.
  My white folks talked plain to me. Dey said real sad like: "Charlie you
is bin a dependence but now you kin go effen you is so desirous. But effen
you wants to stay wid us you kin sharecrop. Dare is a house fur you, en
wood to keep you warm, en a mule to work. We aint got much cash, but
we is got land and you can allus count on havin' plenty ǝ' vittals. Do jist es
you please. When I looked at my marse en knowed he needed me, I
pleased to stay. My marster show nevah forced me to do nary thing.[12]

The editors of the WPA interviews warn us that the opinions must be
read with great caution, since most of the interviews were conducted by
white women upon whose goodwill the elderly blacks often depended
for their very survival. It was likely that the interviewers talked with
"good" elderly blacks in preference to "bad" ones, and the opinions
rendered may have been raised to a formula response known to please
southern whites.[13] Another ex-slave in the district also remembered
according to formula: "After de wah was ober in de Yankees cleared out
all us, what could, went to de white folks we knowed. Dey didn't hab
no money to pay us so we sharecropped. It wuzn't much different from
slavery. We lived in quarters, used de white folks hosses en ploughs en

helped rai'se our own food."[14] Ex-slave Isaac Stier told Mrs. Moore how "All de cullud folks what lived to git back home took to de lan' again. If dey marsters was dead dey went to his friend's an' offered to put in a share crop."[15] Such was the legend.

For Schlomowitz and Reid, the legend largely conforms to their preconceived theory of what ought to have happened in a free market setting. Finding no evidence of planter combinations to coerce labor, freedmen unions, or government intervention, Shlomowitz concludes that the "types of contracts chosen to reward freedmen for their labor were largely determined by market forces."[16] With no constraints except the economic context, the invisible hand of the market resulted in a system for the production and distribution of cotton known as sharecropping. But the free market view is as ahistorical as Percy's traditional view. In both cases, a preconceived and fully developed model, view of the world, or theory, if you wish, has been imposed upon the evidence in such a way as to render the generalizations useless for an understanding of the historical process at work. In the Natchez District, contrary to the traditional and free market views, blacks resisted a return to gang wages, looked to shares as the best means of protecting their existing interest in the crops, and hoped to eventually farm on their own as landowners. Although they acted in their self-interest, it would be wrong to think that freedmen came to sharecropping primarily as individuals interested in making the best economic deal possible in hopes of maximizing incomes. They were interested in maximizing incomes but in the proper setting, one that involved a deeply felt historical constraint: the commitment to family farming free of white supervision.

Sharecropping might not have emerged in the Natchez District but for the role played by the U.S. Army and the Freedmen's Bureau. It is not that these government agencies initiated or even promoted sharecropping. Rather, army and bureau policy enabled district freedmen to resist gang labor and planter determination of their working conditions. Bureau agents hoped to create a setting in which freedmen could sell their labor to the highest bidders, move at will in search of the best opportunities, work as free men for a fair day's pay, and ascend the agricultural ladder to landownership. To reach this goal, the bureau often sided with planters in the enforcement of contracts and the disciplining of labor. But government policy recognized that neither the forces of tradition nor the free market served the best interests of the former slaves. From the time of the Emancipation Proclamation to the last days of the Freedmen's Bureau, it was the announced policy of the relevant government agencies to protect and train the freedmen until they could function on their own. If freedmen could not be set up as free farmers, then they should be protected as free laborers. It was in this context that district freedmen, determined not to become closely supervised workers, used their freedom to become sharecroppers.

Once sharecropping was initiated, the Natchez District was never the same again. The system drastically altered the functioning of all its participants in a way that perpetuated its existence as the dominant mode of production in district agriculture. Although a great many district plantations remained in the hands of their antebellum owners, these landlords were no longer planters in the sense they had been in slavery. Postbellum landowners were unable to coerce labor in quite the same way as before the war. It is not that sharecropping found freedmen managing themselves on their own farms, owned or rented. This was seldom the case. Most commonly, the crucial decisions of crop-making were made and enforced by landlords and merchant suppliers of credit, but the freedmen were free of daily and constant supervision. This lack of supervision made all the difference in the world. For most planters, their new status reduced them to managers of properties, often leased out to merchants, or, if they continued to take an active role in the plantation, they were now businessmen constantly faced with the loss of their labor supply.[17]

In this setting, planters generally abandoned all attempts to recreate the antebellum system of strictly disciplined and closely supervised labor. For one thing, freedmen were free to move, and few would have tolerated such close supervision. For another, the emergence of the merchant as the chief source of supply for the freedmen, as the chief creditor for the planter-landlord, and as the chief lessee of plantation lands permanently settled the system of sharecropping upon the district. The role of the merchant in the institutionalization of sharecropping may not be underestimated. Once the merchant entered the picture, there was little chance for the reemergence of fixed wages or gang labor conditions. The matter was simple enough: a return to a system of close supervision was costly and might prove disastrous if prices should fall. More importantly, with the merchant in control of the crop either as holder of a first lien for supplies advanced or as a landlord holding a lien for rent, there was little economic incentive for these "new" southern businessmen to closely supervise their tenants.

There is yet another, perhaps more fundamental, reason for the persistence of sharecropping in the district. Much of the literature on slavery assumes that the antebellum master was a highly efficient manager of his unwilling and unskilled slaves. They were adept, we are told, at getting from their crude laborers effective and productive work by means of coercion and good management.[18] In the Natchez District, however, the relative efficiency of freedmen as sharecroppers casts doubt upon the notion that southern slave masters were effective managers. Here the productivity of labor in sharecropping, which was neither coerced nor intensely managed, compared favorably with the productivity of labor in slavery. In view of the evidence, it makes as much sense, then, to argue that slave masters had mined the land as to say that they had managed it. Or we could say that slave masters had enjoyed the "luck" of a favorable market for their cotton and their slaves.[19] Sharecropping, being equal to slavery insofar as the relative efficiency of labor was concerned, provided few managerial reasons for a return to an organization of labor similar to slavery.

Hence, the origins of sharecropping in the Natchez District may not best be understood as a freely arrived at bargain between the parties involved. Nor is it useful to think of the period when sharecropping began as an era dominated either by the forces of custom or by the free market. Perhaps most misleading is the attempt to discover or identify some particular mechanism by which sharecropping may be explained. This purpose tends to predetermine the conclusion even before the research is done. There may have been no single factor easily defined as the springboard, nor even an arrangement in which all of the parts came together to form the system. A less mechanistic approach, at least for the Natchez District, suggests that the crucial determining factor in the coming of sharecropping was the freedman's insistence upon it, but not even this explanation is enough unless it is seen in relation to the role of other dynamic forces at play.

Another important consideration is that, in time, sharecropping became the means by which the district's post-Reconstruction economy was held together. In a relatively short time, the problems of production and distribution were settled or contained within its framework. A legal system soon emerged which defined the freedman as a laborer rather than as a freeholding tenant or a land-attached peasant.[20] This development essentially settled the issue of the social control of labor. With few work alternatives available, the legal control of the freedman's crop and food supply enabled his landlords and creditors to effectively dominate the management of labor on the farms. All of the crucial crop decisions were made by someone other than the tenants. It was because of their legal control of the crop that landlords and merchants were able to mold the sharecropping contracts to their interests in keeping the freedmen at work with minimum supervision costs, in keeping the sharecroppers in cotton, and in keeping the district's agricultural workers dependent, impoverished, and largely underemployed. It was in this sense that sharecropping was not much different from slavery: "I kin see de colored folks aint made much ob dey freedom. Day is all in debt en chained down to something same ez us slaves was."[21]

But sharecropping was not the same as slavery. Freedmen lived and worked in families, were relatively unsupervised, and were generally free to move, although the risks of such movement were often overwhelmingly great. As late as the 1930s, few district "croppers," as they were called, half and quarter tenants alike, were closely supervised in their daily work. The general practice found landlords, or their agents, riding the fields after planting (to see that enough acreage was being cultivated to make the rent) and several times during the season. At cotton-picking time, wage hands might be brought in at the tenant's expense and supervised in gangs. In some cases, half tenants were worked by the bell "from can to can't" under close supervision, but this was not the general practice. And even the closely supervised half tenants were able to move at year's end.[22] Some obviously were kept on the plantations by a combination of physical force and legal constraints, but landlords and merchants well understood how quickly their tenants would leave if mistreated. More effective than physical compulsion was the practice of keeping the tenants in debt to the store. Its nearly universal practice amounted to a kind of entrapment that may not have kept tenants at work on one place for very long, but at least meant that there was little real advantage to leaving. But still they moved: "Every year you see dem, whites too, packing up on an ol' wagon an moving somewhere else. Des man done stole from dem an' dey goin' tuh anuthur—tuh let him steal frum dem!"[23]

What began as a desperate attempt to break away from slavery ended in something far different from what the freedmen had hoped to achieve: "We just changed a marster for a boss." But when masters became bosses and slaves became sharecroppers, much had changed. That sharecropping became in a short time a system dominated by the district's racist tradition as well as by the so-called forces of the market meant that its bosses were patriarchs and the workers little more than an impoverished caste and class of ruthlessly controlled dependents. Yet, because sharecropping began in a confrontation born of freedom, its black members were never quite reduced to that state of dependency implied by the words "good and faithful-labor."

NOTES

1. See William Alexander Percy, *Lanterns on the Levee: Recollections of a Planter's Son* (New York: Alfred A. Knopf, 1941); Joseph D. Reid, Jr., "The Evaluation and Implications of Southern Tenancy," *Agricultural History* 53 (July 1979), pp. 153-69.

2. Percy, *Lanterns on the Levee*, pp. 275-76.

3. Ibid., p. 282.

4. Shlomowitz, "The Origins of Southern Sharecropping," *Agricultural History* 53 (July 1979), pp. 557-75.

5. Ibid.

6. Ibid.

7. Percy, *Lanterns on the Levee*, p. 280.

8. Reid, "The Evaluation and Implications of Southern Tenancy," p. 167.

9. Ibid.

10. Ibid.

11. Ibid.

12. George P. Rawick, ed., *The American Slave: A Composite Autobiography* (Westport, Conn.: Greenwood Press, 1977), Supplement Series 1, 8: Mississippi Narr., pp. 565-66.

13. Ibid., 6: xxi-xvi.

14. Ibid., 8: 1347.

15. Ibid., 10: 2054.

16. Shlomowitz, "The Origins of Southern Sharecropping," p.560.

17. See Kenneth S. Greenberg, "The Civil War and the Redistribution of Land: Adams County, Mississippi, 1860-1870," *Agricultural History* 52 (April 1978), pp. 292-307.

18. Reid, "The Evaluation and Implications of Southern Tenancy," pp. 163-66.

19. Ibid., pp. 158-62.

20. See Harold D. Woodman, "Post-Civil War Southern Agriculture and the Law," *Agricultural History* 53 (January 1979), pp. 319-37.

21. Rawick, ed., *American Slave*, 7: Mississippi Narr., p. 572; and Pete Daniel, "The Metamorphosis of Slavery, 1865-1900," *Journal of American History* 66 (June 1979), pp. 88-99.

22. Allison Davis, Burleigh B. Gardner, and Mary R. Gardner, *Deep South: A Social Anthropological Study of Caste and Class* (Chicago: University of Chicago Press, 1941), pp. 286, 329-42.

23. Ibid., p. 341.

_____ APPENDIX A _____
Basis for Estimating Grain Needs, Natchez District, 1860

The grain estimates found in Table 7 are drawn largely from the calculations made by Robert E. Gallman, "Self-Sufficiency in the Cotton Economy of the Antebellum South," and Raymond C. Battalio and John Kagel, "The Structure of Antebellum Southern Agriculture: South Carolina, A Case Study." Both articles have been published in William N. Parker, ed., _The Structure of the Cotton Economy of the Antebellum South_ (Washington, D.C.: Agricultural History Society, 1970), pp. 5-38. In almost every case, I have utilized their maximum estimates in order to guard against the possibilities of underestimating the reality. Where they differed, I usually selected the larger estimate. For a detailed discussion of the difficulties in estimating grain needs, see Roger L. Ransom and Richard Sutch, _One Kind of Freedom: The Economic Consequences of Emancipation_ (Cambridge, England: Cambridge Univerity Press, 1977), pp. 244-53, 363-64.

Appendix Table 1
Consumption Needs (Annual), Slave Plantation, 1860

	Corn (bushels)
Adults, 16-65 years	18
Secondary hands, 10-15 years	10
Children and elderly	6
Cows	9
Other cattle	3
Swine	½ at 7, ½ at 2
Mules, oxen, horses	38
Unknown young animals	½ amount consumed by other cattle
Sheep	0.5
Seeds	5 percent of total produced

SOURCE: Robert E. Gallman, "Self-Sufficiency in the Cotton Economy of the Antebellum South,"and Raymond C. Battalio and John Kagel, "The Structure of Antebellum Southern Agriculture: South Carolina, A Case Study." Both articles have been published in William N. Parker, ed., *The Structure of the Cotton Economy of the Antebellum South* (Washington, D.C.: Agricultural History Society, 1970), pp 5-38.

Basis for Estimating Production Data, Natchez District, 1860

The production data contained in Appendix Tables 1 and 2 were compiled from the Manuscript Census for the Natchez District. Each farm or plantation was evaluated and categorized according to the size of its slave force. Not every farm entry was included. In some cases the material was so misleading (unclear, vague, and so on) as to be useless for study. Once the labor force was determined, each farm was then studied in terms of the various factors of production employed and the crops harvested.

Appendix Table 2
Input-Output Data, Adams County, 1860

Size of Farm Measured in Slaves	Size of Sample (pct.)	Number of Prime Hands	Number of Children	Work Animals	Value of Improved Acre (average)	Value of Tools	Cotton	Corn
1-5	25	18	2	34	$ 9.78	$ 30	44	1,300
6-10	34	48	8	66	19.78	430	154	3,200
11-20	40	160	59	182	52.30	1,627	583	15,700
21-30	80	327	89	334	17.88	2,850	1,006	17,300
31-50	87	1,175	273	1,006	23.79	10,156	4,087	53,600
51-80	91	1,572	362	1,301	22.08	13,180	4,763	66,408
81-100	94	1,102	269	887	19.08	7,970	3,632	42,100
101-150	92	1,011	258	626	10.06	7,000	3,239	44,500
151-200	87	1,007	207	504	17.86	4,130	2,695	15,400
201-250	100	167	37	189	42.65	3,000	700	15,000
251-300	0							
301 +	100	872	203	517	31.24	14,000	3,185	39,000

SOURCE: Manuscript Census.

Appendix Table 3
Input-Output Data, Concordia Parish, 1860

Size of Farm Measured in Slaves	Size of Sample (pct.)	Number of Prime Hands	Number of Children	Work Animals	Value of Improved Acre (average)	Value of Tools	Cotton	Corn
1-5	50	34	4	46	$ 11.62	$ 745	79	3,925
6-10	76	112	32	129	78.18	3,965	440	7,400
11-20	88	301	97	344	81.66	42,050	2,137	23,785
21-30	89	163	64	163	100.82	20,650	614	12,000
31-50	81	601	150	506	99.41	56,525	3,551	26,600
51-80	97	1,425	478	1,259	91.91	134,635	9,889	93,355
81-100	100	1,046	345	869	107.89	89,370	8,445	66,200
101-150	96	2,064	764	1,510	127.85	192,633	17,179	110,185
151-200	100	1,821	631	1,184	118.59	179,459	13,443	109,635
201-250	100	674	204	409	118.11	73,700	4,223	19,660
251-300	100	196	63	119	123.40	45,000	1,750	10,000
301 +								

SOURCE: Manuscript Census.

203

APPENDIX **C.** ────────
Basis for Estimating the Labor Status of Freedmen, Natchez District, 1880

Approximately 22,000 blacks lived in the Natchez District in 1880, of which only about six hundred to eight hundred were themselves, or the family of, owner-operators. The procedure for estimating the number of blacks in the various status groups was to establish the number of farm entries in the manuscript census of agriculture, sum the total unknown with the known in regard to race, and multiply by the average number of people found in each household. (Appendix Table 4.)

Appendix Table 4
Farming Population by Status, Adams County and Concordia Parish, 1880

	Adams	Concordia
Total farm entries	1,911	1,318
Black owners	119	4
Black tenants	156	73
Black sharecroppers	853	857
White owners	108	71
White tenants	22	69
White sharecroppers	43	19
Unknown owners	129	25
Unknown tenants	95	48
Unknown sharecroppers	138	152

NOTE: The average number of people in each family with the old, young, and nonfamily members included was five (5).

SOURCES: U.S. Census (1880), Manuscript Population and Agricultural Schedules, Adams County, Mississippi, and Concordia Parish, Louisiana.

Included in the agricultural schedules of the manuscript census are data on wages paid and the number of weeks for which wages were paid on each farm listed. Although the data are difficult to use, since they fail to indicate the actual number of people employed or the wages paid per worker, it is possible to work with the information given to estimate the number of full-time hands employed on each farm. The procedure is to assume that any farmer who hired labor for at least sixteen weeks out of the year was dealing with full-time hired hands. (This assumption seems realistic since the average farmer employing hands over sixteen weeks paid wages forty-four weeks, as an average, out of the year.) By then dividing the total amount of wages paid by the above farmers by the approximate yearly wages earned by full-time hands—about $120 a year—we arrive at the total number of people employable as full-time hands by the monies spent. Although this figure overestimates the number of people actually employed as hired hands, since farmers seldom hired their hands for a full twelve-month period at the rate of $10 a month, it nevertheless provides some idea of the maximum number of people so employed in the Natchez District. It should be noted, too, that the $10 a month figure is a rather conservative estimate, if the total wage bill cited on each farm included the value of rations and clothing which some farmers clearly designated as part of the freedman's wages. In any case, the amount is conservative in view of the various estimates available which range from $16 to $20 a month.

If we assume that the remaining monies listed in the wages paid columns, usually monies paid by sharecroppers and tenants, were paid to pickers at the estimated rate of $1 per day, we may then come close to estimating the actual number of people employed as pickers (Appendix Tables 5 and 6). Not every picker, of course, made the high of $1 per day, since the pickers were usually paid on the basis of the pounds picked, but the estimates published by Edward Young in 1874 suggest that the amounts usually ranged from 50 cents to the high of $1. By accepting the high amount, we avoid the risk of overestimating the number of people possibly employed as pickers on a cash wage basis. (For wage estimates, see U.S. Congress, House, Report of the Commissioner of Agriculture, 38th Cong., 2d Sess., 1864, *House Exec. Doc.* 12, pp. 88-91; "Labor in America and Europe," by Edward Young, 44th Cong., 1st Sess., 1876, *House Exec. Doc.* 21, pp. 739-44.)

Appendix Table 5
Estimates of Wage Laborers: Full- and Part-Time Workers, 1880

	Adams	Concordia
Number of workers employed as full-time hands paid at rate of $120 a Year	264	514
Number of workers employed for six weeks at $1 per day	663	4,020

SOURCES: U.S. Census (1880), Manuscript Population and Agricultural Schedules, Adams County, Mississippi, and Concordia Parish, Louisiana.

Appendix Table 6
Estimates of Total Number of Freedmen (Men, Women, and Children) in each Status Group, Adams County and Concordia Parish, 1880 (Based on the calculation of five people in each family)

	Adams	Concordia
Black owners	595	20
Black tenants	780	365
Black sharecroppers	4,265	4,385
Unknown tenants	475	125
Unknown sharecroppers	1,931	760
Full-time wage hands[a]	792	1,542
Total	8,838	7,197
Cash wage pickers	663	4,020

[a]Unlike the sharecroppers and tenants, the full-time wage hands probably had fewer people living with them who were not family members. I have thus calculated their numbers on the basis of three people per family.

SOURCES: U.S. Census (1880), Manuscript Population and Agricultural Schedules, Adams County, Mississippi, and Concordia Parish, Louisiana.

SELECTED BIBLIOGRAPHY

MANUSCRIPTS

Baton Rouge, Louisiana. Louisiana State University Library. Lemuel Parker
Conner Family Papers.
———. William Dunbar Papers.
———. Duncan Family Papers.
———. J. A. Gillespie Papers.
———. Good Hope Plantation Papers.
———. William Lovell Papers.
———. William Newton Mercer Papers.
———. William J. Minor Papers.
———. George W. Montgomery Papers.
———. Joseph Addison Montgomery Papers.
———. Police Jury Minutes for Concordia Parish, Louisiana (1861-1877).
———. J. D. Shields Papers.
Baton Rouge, Louisiana. State Comptroller's Office. Manuscript Tax Rolls,
(1861-1896), for Concordia Parish, Louisiana.
Cambridge, Massachusetts. Baker Library, Harvard University. Mercantile Agency
Credit Ledgers, R. G. Dun and Company. (Used and cited by permission
of Harvard University.)
Chapel Hill, North Carolina. Southern Historical Collections, University of
North Carolina Library. Thomas Butler King Papers.
———. Quitman Family Papers.
———. U.S. Census (1860), Manuscript Agricultural, Population and Slave
Schedules for Adams County, Mississippi, and Concordia Parish,
Louisiana.

208

Selected Bibliography

————. U.S. Census (1870), Manuscript Population and Agricultural Schedules for Adams County, Mississippi, and Concordia Parish, Louisiana.
————. U.S. Census (1880), Manuscript Population and Agricultural Schedules for Adams County, Mississippi, and Concordia Parish, Louisiana.
Jackson, Mississippi. Department of Archives and History. Manuscript Tax Rolls (1818-1861), for Adams County, Mississippi.
Natchez, Mississippi. Office of Records. Chancery Records (1869-1875), for Adams County, Mississippi.
————. Liens and Mortgage Records (1850-1890), for Adams County, Mississippi.
————. Probate Records (1870-1880), for Adams County, Mississippi.
Vidalia, Louisiana. Office of Records. Liens and Mortgage Records (1850-1890), for Concordia Parish, Louisiana.
Washington, D.C. National Archives. Record Groups 363 and 366. Records of the Adjutant General—Colored Troops Division.
————. Record Group 94. Records of the Adjutant General's Office, General's Papers.
————. Record Groups 105 and 752. Records of the Bureau of Refugees, Freedmen, and Abandoned Lands.
————. Record Group 366. Treasury Department, Civil War Special Agency Records.

PUBLIC DOCUMENTS

Harper, L. *Report on the Geology and Agriculture of the State of Mississippi.* Mississippi: E. Barksdale, State Printer, 1857.
Hilgard, E. *Report on the Geology and Agriculture of the State of Mississippi.* Mississippi: E. Barksdale, State Printer, 1860.
Louisiana. *Annals of the Acts Passed by the General Assembly of the State of Louisiana* (1865-1890).
————. *Reports of Cases Adjudged in the Supreme Courts of Louisiana* (1865-1900). Mississippi. *Laws of Mississippi* (1865-1900).
————. *Reports of Cases Adjudged in the Supreme Courts of Mississippi* (1865-1900).
U.S. Congress. House, Circulars, General Orders, and Reports of the Freedmen's Bureau. *House Executive Document.* 70, 39th Cong., 1st Sess., 1866.
————. *House.* Labor in America and Europe, by Edward Young. *House Executive Document.* 21, 44th Cong., 1st Sess., 1876.
————. Report of the Commissioner of Agriculture. *House Executive Document.* 12, 38th Cong., 2d Sess., 1864.
————. Report of the Joint Committee on Reconstruction. *H. Rept.* 30, 39th Cong., 1st Sess., 1866.

————. *Senate*. Reports of the Assistant Commissioners of the Freedmen's Bureau to General O. O. Howard. *Senate Executive Document*. 6, 39th Cong., 2d Sess., 1866.

————. Report to the War Department on the Levees of Mississippi. *S. Rept.* 126, 40th Cong., 1st Sess., 1866.

U.S. Department of Commerce. *Bureau of the Census. Eighth Census of the United States: 1860.* Agriculture. Vol. 3.

————. *Eighth Census of the United States: 1860. Population.* Vol. 1.

————. *Tenth Census of the United States: 1880. Agriculture.* Vol. 3.

————. *Tenth Census of the United States: 1880. Population.* Vol. 1

Wailes, B.L.C. *Report on the Agriculture of the State of Mississippi Embracing a Sketch of the Social and Natural History of the State.* Mississippi: E. Barksdale, State Printer, 1854.

The War of Rebellion: A Compilation of the Official Records of the Union and Confederate Armies. 128 vols. Washington, D.C.: U.S. Government Printing Office, 1880-1901.

NEWSPAPERS AND PERIODICALS

American Agriculturist, 1849-1868.

The American Farmer, 1868.

Courier (Natchez), 1850-1867.

Cultivator and Country Gentleman, 1865-1890.

Daily Crescent (New Orleans), 1866.

Democrat (Natchez), 1865-1872.

Intelligencer (Concordia), 1859, 1867, 1869.

Southern Cultivator, 1867-1870.

Southern Planter, 1867-1868.

Times (New York), 1863-1900.

BOOKS

Atherton, Lewis. *The Southern Country Store.* New York: Greenwood Press, 1968. (First published in 1949.)

Banks, Enoch. *The Agrarian Revolution in Georgia.* New York: Columbia University Press, 1905.

Bass, Herbert J., ed. *The State of American History.* Chicago: Quadrangle Books, 1970.

Bentley, George R. *A History of the Freedmen's Bureau.* New York: Octagon Books, 1970. (First published in 1955.)

210 *Selected Bibliography*

Billings, Dwight B., Jr. *Planters and the Making of a "New South": Class, Politics, and Development in North Carolina, 1865-1900.* Chapel Hill, N.C.: University of North Carolina Press, 1979.

Bittersworth, John K. *Confederate Mississippi: The People and Politics of a Cotton State.* Baton Rouge, La.: Louisiana State University Press, 1943.

Blassingame, John W. *The Slave Community: Plantation Life in the Ante-bellum South.* New York: Oxford University Press, 1972.

Brandfon, Robert L. *Cotton Kingdom of the New South.* Cambridge, Mass.: Harvard University Press, 1964.

Braverman, Harry. *Labor and Monopoly Capital: The Degradation of Work in the Twentieth Century.* New York: Monthly Review Press, 1974.

Brooks, Robert P. *The Agrarian Revolution in Georgia, 1865-1912.* Madison, Wis.: University of Wisconsin Press, 1914.

Callender, Gray S., ed. *Economic History of the United States.* New York: A. M. Kelley, 1965.

Chayanov, A. V. *The Theory of Peasant Economy.* Homewood, Ill.: American Economic Association, 1966.

Cheung, Steven N.S. *The Theory of Share Tenancy with Special Application to Asian Agriculture and the First Phase of Taiwan Land Reform.* Chicago: University of Chicago Press, 1969.

Clark, Thomas Dionysius. *Pills, Petticoats, and Plows: The Southern Country Store.* New York: Bobbs-Merrill Co., 1944.

Courtenay, Philip P. *Plantation Agriculture.* New York: Praeger, 1966.

David, Paul A., Herbert G. Gutman, Richard Sutch, Peter Temin, and Gavin Wright, eds. *Reckoning with Slavery: A Critical Study in the Quantitative History of American Negro Slavery.* New York: Oxford University Press, 1976.

Davis, Allison, Burleigh B. Gardner, and Mary R. Gardner. *Deep South: A Social Anthropological Study of Caste and Class.* Chicago: University of Chicago Press, 1941.

DeCanio, Stephen J. *Agriculture in the Postbellum South: The Economics of Production and Supply.* Cambridge, Mass.: MIT Press, 1974.

Eaton, John, *Grant, Lincoln, and the Freedmen.* New York: Longmans, Green & Co., 1907.

Elkins, Stanley. *Slavery.* Chicago: University of Chicago Press, 1958.

Faulkner, William. *Intruder in the Dust.* New York: Random House, 1948.

———. *The Unvanquished.* New York: Random House, 1938.

Fogel, Robert William, and Stanley Engerman. *Time on the Cross: The Economics of American Negro Slavery.* Boston: Little, Brown & Co., 1974.

Foner, Eric. *Politics and Ideology in the Age of the Civil War.* New York: Oxford University Press, 1980.

Fredrickson, George M. *The Black Image in the White Mind: The Debate on Afro-American Character and Destiny, 1817-1914.* New York: Harper & Row. 1971.

Garner, James W. *Reconstruction in Mississippi.* Baton Rouge, La.: Louisiana State University Press, 1968. (First Published in 1901.)

Gates, Paul W. *The Farmer's Age: Agriculture, 1815-1860.* New York: Holt, Rinehart & Winston, 1960.

Genovese, Eugene D. *The Political Economy of Slavery: Studies in the Economy of Slavery: Studies in the Economy and Society of the Slave South.* New York: Pantheon Books, 1961.

———. *Roll, Jordan, Roll: The World the Slaves Made.* New York: Pantheon Books, 1974.

———. *The World the Slaveholders Made.* New York: Vintage Books Edition, 1971.

Gerteis, Louis S. *From Contraband to Freedman: Federal Policy Towards Southern Blacks, 1861-1865.* Westport, Conn.: Greenwood Press, 1973.

Grant, Ulysses S. *Personal Memoirs of U. S. Grant.* 2 vols. New York: C. L. Webster & Co., 1896.

Gray, Lewis Cecil. History of Agriculture in the Southern United States to 1800. 2 vols. Washington, D.C.: Carnegie Institute of Washington, 1933.

Gutman, Herbert G. *The Black Family in Slavery and Freedom, 1750-1925.* New York: Pantheon Books, 1976.

———. *Work, Culture and Society in Industrializing America.* New York: Vintage Books, 1977.

Hair, William Ivy. *Bourbonism and Agrarian Protest: Louisiana Politics, 1877-1900.* Baton Rouge, La.: Louisiana State University Press, 1969.

Harris, William C. *Presidential Reconstruction in Mississippi.* Baton Rouge, La.: Louisiana State University Press, 1967.

Hermann, Janet Sharp. *The Pursuit of a Dream.* New York: Oxford University Press, 1981.

Higgs, Robert. *Competition and Coercion: Blacks in the American Economy, 1865-1914.* New York: Cambridge University Press, 1977.

Howard, Oliver O. *Autobiography of Oliver Otis Howard.* 2 vols. New York: Baker & Taylor Co., 1907.

James, D. Clayton. *Antebellum Natchez.* Baton Rouge, La.: Louisiana State University Press, 1968.

Johnson, Charles S. *The Collapse of Cotton Tenancy.* Chapel Hill, N.C.: University of North Carolina Press, 1933.

———. *Shadow of the Plantation.* Chicago: University of Chicago Press, 1934.

Knox, Thomas W. *Camp-Fire and Cotton-Field*. New York: Blelock & Co., 1865.

Kolchin, Peter. *First Freedom: The Response of Alabama's Blacks to Emancipation and Reconstruction*. Westport, Conn.: Greenwood Press, 1972.

Litwack, Leon. *Been in the Storm So Long: The Aftermath of Slavery*. New York: Alfred A. Knopf, 1979.

Loring, F. W., and C. F. Atkinson. *Cotton Culture and the South Considered with Reference to Emigration*. Boston: A. Williams Co., 1869.

Luraghi, Raimondo. *The Rise and Fall of the Plantation South*. New York: New Viewpoints, 1978.

McFeely, William S. *Yankee Stepfather: A Study of General O. O. Howard and the Freedmen's Bureau*. New Haven, Conn: Yale University Press, 1968.

McPherson, James M. *The Negro's Civil War*. New York: Random House, 1965.

Magdol, Edward. *A Right to the Land: Essays on the Freedmen's Community*. Westport, Conn.: Greenwood Press, 1977.

Mandle, Jay R. *The Roots of Black Poverty: The Southern Plantation Economy After the Civil War*. Durham, N.C.: Duke University Press, 1978.

Mangum, Charles S., Jr. *The Legal Status of the Tenant Farmer in the Southwest*. Chapel Hill, N.C.: University of North Carolina Press, 1952.

Menn, Joseph Karl. *The Large Slaveholders of Louisiana—1860*. New Orleans: Pelican Publishing Co., 1964.

Moore, John Hebron. *Agriculture in Ante-bellum Mississippi*. New York: Octagon Books, 1971. (First published in 1958.)

Myrdal, Gunnar. *The American Dilemma: The Negro Problem and Modern Democracy*. 2 vols. New York: Harper & Row, 1944.

Nielson, A. E. *Production Credit for Southern Cotton Growers*. Morningside Heights, N.Y.: King's Crown Press, 1946.

Nieman, Donald G. *To Set the Law in Motion: The Freedmen's Bureau and the Legal Rights of Blacks, 1865-1868*. Millwood, N.Y.: KTO Press, 1979.

Nordhoff, Charles. *The Cotton States in the Spring and Summer of 1875*. New York: Burt Franklin, 1876.

Novak, Daniel. *The Wheel of Servitude: Black Forced Labor After Slavery*. Lexington, Ky.: University Press of Kentucky, 1978.

Nutt, Merle C. *The Nutt Family Through the Years, 1635-1973*. Phoenix, Ariz.: Merle C. Nutt, 1973.

O'Gorman, Ned. *The Children Are Dying*. New York: Signet, 1978.

Olmsted, Frederick Law. *The Cotton Kingdom: A Traveller's Observations on Cotton and Slavery in the American Slave States*. New York: Mason Brothers, 1861.

Oubre, Claude F. *Forty Acres and a Mule: The Freedmen's Bureau and Black Land Ownership.* Baton Rouge, La.: Louisiana State University Press, 1978.

Otken, Charles H. *The Ills of the South, or Related Causes Hostile to the General Prosperity of the Southern People.* New York: G. P. Putnam's Sons, 1894.

Owsley, Frank Lawrence. *Plain Folk of the Old South.* Baton Rouge, La.: Louisiana State University Press, 1949.

Parker, William N., ed. *The Structure of the Cotton Economy of the Antebellum South.* Washington, D.C.: Agricultural History Society, 1970.

Painter, Nell I. *Exodusters: Black Migration to Kansas after Reconstruction.* New York: W. W. Norton & Co., 1976.

Percy, William Alexander. *Lanterns on the Levee: Recollections of a Planter's Son.* New York: Alfred A. Knopf, 1941.

Phillips, U. B. *American Negro Slavery.* Baton Rouge, La.: Louisiana State University Press, 1966. (First published in 1918.)

―――. *Life and Labor in the Old South.* Boston: Little, Brown & Co., 1929.

Pierce, Paul Skuls. *The Freedmen's Bureau: A Chapter in the History of Reconstruction.* Iowa City, Iowa: University of Iowa, 1944.

Postell, William Dosite. *The Health of Slaves on Southern Plantations.* Baton Rouge, La.: Louisiana State University Press, 1951.

Powdermaker, Hortense. *After Freedom: A Cultural Study of the Deep South.* New York: Atheneum Press, 1968.

Powell, Lawrence N. *New Masters: Northern Planters During the Civil War and Reconstruction.* New Haven, Conn.: Yale University Press, 1980.

Range, Willard. *Century of Georgia Agriculture 1850-1950.* Athens, Ga.: University of Georgia Press, 1954.

Ransom, Roger L, and Richard Sutch. *One Kind of Freedom: The Economic Consequences of Emancipation.* Cambridge, England: Cambridge University Press, 1977.

Raper, Arthur. *Preface to Peasantry: A Tale of Two Black Belt Counties.* Chapel Hill, N.C.: University of North Carolina Press, 1935; New York: Atheneum Press, 1968.

Rawick, George P., ed. *The American Slave: A Composite Autobiography.* Supplement Series 1, 6, 7, 8, 10, Mississippi Narr. Westport, Conn.: Greenwood Press, 1977.

―――. *From Sundown to Sunup: The Making of the Black Community.* Westport, Conn.: Greenwood Press, 1972.

Reid, Whitelow. *After the War: A Southern Tour, May 1, 1865 to May 1, 1866.* New York: Moore, Wilstock & Baldwin, 1866.

214   *Selected Bibliography*

Ripley, C. Peter. *Slaves and Freedmen in Civil War Louisiana.* Baton Rouge, La.: Louisiana State University Press, 1976.

Roark, James L. *Masters Without Slaves: Southern Planters in the Civil War and Reconstruction.* New York: W. W. Norton & Co., 1977.

Rogers, William Warren. *The One-Gallused Rebellion: Agrarianism in Alabama, 1865-1896.* Baton Rouge, La.: Louisiana State University Press, 1970.

Rose, Willie Lee. *Rehearsal for Reconstruction: The Port Royal Experiment.* New York: Vintage Books, 1964.

Rosengarten, Theodore. *All God's Dangers: The Life and Times of Nate Shaw* New York: Avon Books, 1975.

Rubin, Morton. *Plantation County.* Chapel Hill, N.C.: University of North Carolina Press, 1951.

Scarborough, William Kauffman. *The Overseer: Plantation Management in the Old South.* Baton Rouge, La.: Louisiana State University Press, 1966.

Schultz, T. H. *Transforming Traditional Agriculture.* New Haven, Conn.: Yale University Press, 1964.

Schurz, Carl. *The Reminiscences of Carl Schurz.* 2 vols. New York: McClure Co., 1908.

Schwartz, Michael. *Radical Protest and Social Structure: The Southern Farmers Alliance and Cotton Tenancy.* New York: Academic Press, 1976.

Sefton, James E. *The United States Army and Reconstruction, 1865-1877.* Baton Rouge, La.: Louisiana State University Press, 1967.

Shannon, Fred A. *The Farmer's Last Frontier: Agriculture, 1860-1897.* New York: Harper Torchbooks Edition, 1968.

Shugg, Roger W. *Origins of Class Struggle in Louisiana: A Social History of White Farmers and Laborers During Slavery and After, 1840-1875.* Baton Rouge, La.: Louisiana State University Press, 1939.

Somers, Robert. *The Southern States Since the War, 1870-1871.* University, Ala.: University of Alabama Press, 1965. (First published in 1871.)

Stampp, Kenneth Milton. *The Peculiar Institution: Slavery in the Ante-bellum South.* New York: Alfred A. Knopf, 1956.

Sydnor, Charles S. *Slavery in Mississippi.* Baton Rouge, La.: Louisiana State University Press, 1966. (First published in 1933.)

Thompson, E. P. *The Making of the English Working Class.* New York: Vintage Books, 1963.

Thompson, Edgar. *Plantation Societies; Race Relations and the South: The Regimentation of Populations.* Durham, N.C.: Duke University Press, 1975.

Vance, Rupert. *All These People.* Chapel Hill, N.C.: University of North Carolina Press, 1945.

Van Deburg, William L. *The Slave Drivers: Black Agricultural Labor Supervisors in the Antebellum South*. Westport, Conn.: Greenwood Press, 1979.

Weaver, Herbert. *Mississippi Farmers, 1850-1860*. Nashville, Tenn.: Vanderbilt Press, 1945.

Wharton, Vernon L. *The Negro in Mississippi, 1865-1890*. Chapel Hill, N.C.: University of North Carolina Press, 1947.

White, Howard A. *The Freedmen's Bureau in Louisiana*. Baton Rouge, La.: Louisiana State University Press, 1970.

Wiener, Jonathan M. *Social Origins of the New South: Alabama, 1860-1885*. Baton Rouge, La.: Louisiana State University Press, 1978.

Wiley, Bell Irvin. *Southern Negroes, 1861-1865*. New Haven, Conn.: Yale University Press, 1938.

Williamson, Joel. *After Slavery: The Negro in South Carolina During Reconstruction, 1861-1877*. Chapel Hill, N.C.: University of North Carolina Press, 1965.

Woodman, Harold D. *King Cotton and His Retainers: Financing and Marketing the Cotton Crop of the South, 1800-1925*. Lexington, Ky.: University of Kentucky Press, 1968.

Woodward, C. Vann. *Origins of the New South, 1877-1913*. Baton Rouge, La.: Louisiana State University Press, 1951.

Wright, Gavin. *The Political Economy of the Cotton South: Households, Markets, and Wealth in the Nineteenth Century*. New York: W. W. Norton & Co., 1978.

Yeatman, James E. *A Report on the Condition of the Freedmen of the Mississippi Valley*. St. Louis, Mo.: Western Sanitary Commission Reports, 1864.

ARTICLES AND ESSAYS

Abbot, Martin. "Free Land, Free Labor, and Freedmen's Bureau, 1865-1868." *Agricultural History* 30 (October 1956), pp. 150-56.

Allen, C. E. "Greater Agricultural Efficiency for the Black Belt of Alabama." American Academy of Political Science, *Annals* 61 (September 1915), pp. 187-98.

Anderson, George L. "The South and Problems of Post Civil War Finance." *Journal of Southern History* 9 (June 1943), pp. 181-95.

Aufhauser, Keith. "Slavery and Scientific Management." *Journal of Economic History* 32 (December 1973), pp. 811-23.

Auken, Sheldon Van. "A Century of Southern Plantations." *Virginia Magazine of History* 64 (October 1958), pp. 359-71.

Barzel, Yoram. "An Economic Analysis of Slavery." *Journal of Law and Economics* (April 1977), pp. 102-104.

Bigelow, Martha Mitchell. "Freedmen of the Mississippi Valley, 1862-1865." *Civil War History* 8 (March 1962), pp. 38-47.

Breen, T. H. "The Culture of Agriculture: From Tobacco to Wheat in Tidewater, Virginia, 1760-1790." Paper presented to the annual meeting of the American Historical Association, Washington, D.C., December 31, 1981.

Brown, William W., and Morgan O. Reynolds, "Debt Peonage Re-examined." *Journal of Economic History* 33 (December 1973), pp. 862-71.

Buechel, F. A. "Relationships of Landlords to Farming Tenants." *Land Economies* 1 (July 1925), pp. 336-42.

Bull, Jacqueline D. "The General Merchant in the Economic History of the South, 1865-1900." *Journal of Southern History* 13 (February 1952), pp. 37-59.

Calhoun, Robert Dabney. "A History of Concordia Parish, Louisiana." *Louisiana History Quarterly* 15 (January 1932), pp. 44-67; 15 (April 1932), pp. 214-33; 15 (July 1932), pp. 428-52; 15 (October 1932), pp. 618-45; 16 (January 1933), pp. 92-124.

Clark, Thomas D. "The Furnishing and Supply System in Southern Agriculture Since 1865." *Journal of Southern History* 12 (February 1946), pp. 24-44.

Cox, LaWanda. "The American Agricultural Wage Earner, 1865-1900: The Emergence of a Modern Problem." *Agricultural History* 12 (April 1948), pp. 95-114.

———. "The Promise of Land for the Freedmen." *Mississippi Valley Historical Review* 45 (June 1958), pp. 413-40.

———. "Tenancy in the U.S., 1865-1900: A Consideration of the Validity of the Agricultural Ladder Hypothesis." *Agricultural History* 13 (July 1944), pp. 97-105.

Daniel, Pete. "The Metamorphosis of Slavery, 1865-1900." *Journal of American History* 66 (June 1979), p. 88-99.

Dillingham, Pitt. "Land Tenure Among the Negroes." *Yale Review* (August 1896), pp. 4-17.

Edwards, Thomas J. "The Tenant System and Some Changes Since Emancipation." American Academy of Political Science, *Annals* 49 (September 1913), pp. 38-46.

Ely, Richard T. "Tenancy in an Ideal System of Land Ownership." *American Economic Review* 9 (March 1919), pp. 180-212.

———. "The Experts." *The Crisis* (March 1913), pp. 239-40.

Fogel, Robert W., and Stanley L. Engerman. "Explaining the Relative Efficiency of Slave Agriculture in the Antebellum South." *Americana Economic Review* (June 1977), pp. 275-94.

————. "The Relative Efficiency of Slavery: A Comparison of Northern and Southern Agriculture in 1860." *Explorations in Entrepreneurial History* 8 (Spring 1971), pp 354-64.

Gates, Paul W. "Southern Investments in Northern Lands Before the Civil War." *Journal of Southern History* 5 (May 1939), pp. 155-85.

Gray, Lewis Cecil. "Economic Efficiency and Competitive Advantages of Slavery Under the Plantation System." *Agricultural History* 4 (April 1930), pp. 31-47.

Greenberg, Kenneth S. "The Civil War and the Redistribution of Land: Adams County. Mississipi, 1860-1870." *Agricultural History* 52 (April 1978), pp. 292-307.

Griffen, Richard W. "Problems of the Southern Cotton Planter After the Civil War, 1865-1869." *Georgia Historical Quarterly* 38 (June 1955), pp. 102-17.

Hammond, M. B. "The Southern Farmer and the Cotton Question." *Political Science Quarterly* 12 (September 1897), pp. 451-65.

Hany, Louis H. "Farm Credit Problems in a Cotton State." *American Economic Review* 4 (March 1914), pp. 47-67.

Harris, William. "Formulation of the First Mississippi Plan: The Black Code of 1865." *Journal of Mississippi History* 29 (November 1967), pp. 181-89.

Hibbard, Benjamin H. "Tenancy in the Southern States." *Quarterly Journal of Economics* 27 (March 1913), p. 482-96.

Higgs, Robert. "Did Southern Farmers Discriminate?" *Agricultural History* 46 (April 1972), pp. 325-28.

————. "Patterns of Farm Rental in the Georgia Cotton Belt, 1880-1900." *Journal of Economic History* 34 (June 1974), p. 468-82.

————. "Race, Tenure, and Resource Allocation in Southern Agriculture, 1865-1910." *Journal of Economic History* 33 (March 1973), pp. 149-69.

Highsmith, William E. "Louisiana Landholding During War and Reconstruction." *Louisiana Historical Quarterly* 38 (January 1955), pp. 27-38.

Holmes, George K. "The Peons of the South." American Academy of Political Science, *Annals* 4 (March 1893), pp. 265-74.

Laird, William E., and James R. Rinehart. "Deflation, Agriculture, and Southern Development." *Agricultural History* 32 (April 168), pp. 115-25.

Land, Aubrey C. "Economic Behavior in a Planting Society: The Eighteenth Century Chesapeake." *Journal of Southern History* 33 (Autumn 1967), pp. 467-85.

Mendenhall, Marjorie Stratford. "The Rise of Southern Tenancy." *Yale Review* 27 (Autumn 1937), pp. 110-29.

Metzer, Jacob. "Rational Management, Modern Business Practices, and Economics of Scale in the Ante-Bellum Southern Plantations." *Explorations in Economic History* 12 (April 1975), p 123-50.

Neal, Ernst E. "The Place of the Negro Farmer in the Changing Economy of the Cotton South, 1880-1950." Rural Sociology, 15 (March 1950), pp. 30-51.

Parker, William. "The South in the National Economy, 1865-1970," *Southern Economic Journal* 46 (April, 1980), pp. 1019-48.

Phillips, U. B. "Conservation and Progress in the Cotton Belt." *South Atlantic Quarterly* 3 (January 1904), pp. 1-10.

———. "The Decadence of the Plantation System." American Academy of Political Science, *Annals* 35 (January 1910), pp. 37-56.

———. "The Economics of the Plantation." *South Atlantic Quarterly* 2 (July 1903), pp. 231-36.

———. "Plantations with Slave Labor and Free." *American Historical Review* 30 (July 1925), pp. 738-53.

Putnam, George. "Agricultural Credit and the Tenancy Problem." *American Economic Review* 5 (December 1915), pp. 805-15.

Reid, Joseph D. Jr. "The Evaluation and Implications of Southern Tenancy." *Agricultural History* 53 (January 1979), pp. 153-69.

———. "Sharecropping and Agricultural Uncertainty." *Economic Development and Cultural Change* 24 (April 1976), pp. 549-76.

———. "Sharecropping as an Understandable Market Response: The Post-Bellum South." *Journal of Economic History* 33 (March 1973), pp. 106-30.

———. "Sharecropping in History and Theory." *Agricultural History* 49 (April 1975), pp 426-40.

———. "The Theory of Slave Tenancy Revisited—Again." *Journal of Political Economy* 85 (April 1977), pp. 403-7.

Roger, Henry Wade. "The Law in Relation to Crops." *Southern Law Review* 8 (October 1882), pp. 326-53.

Ross, Steven Joseph. "Freed Soil, Freed Labor, Freed Men: John Eaton and the Davis Bend Experiment." *Journal of Southern History* 64 (May 1978), pp. 213-32.

Rothstein, Morton. "The Antebellum South as a Dual Economy: A Tentative Hypothesis." *Agricultural History* 41 (October 1967), pp. 373-82.

Russell, Robert R. "The Effects of Slavery upon Nonslaveholders in the Antebellum South." *Agricultural History* 15 (April 1941), pp 112-26.

Saloutos, Theodore. "Southern Agriculture and the Problems of Readjustment, 1865-1877." *Agricultural History* 30 (April 1956), pp. 58-76.

Schickele, Rainer. "Effects of Tenure Systems on Agricultural Efficiency." *Journal of Farm Economics* 29 (February 1941), pp. 185-207.

Schmitz, Mark D. "Economies of Scale and Farm Size in the Antebellum Sugar Sector." *Journal of Economic History* 37 (December 1977), pp. 959-80.

Schlomowitz, Ralph. "The Origins of Southern Sharecropping." *Agricultural History* 53 (July 1979), pp. 557-75.

Shugg, Roger W. "Survival of the Plantation System in Louisiana." *Journal of Southern History* 3 (August 1937), pp. 311-25.

Sitterson, J. Carlyle. "The Transition from Slave to Free Economy on the William J. Minor Plantations." *Agricultural History* 17 (October 1943), pp 216-24.

Sutch, Richard. and Roger Ransom. "The Ex-Slave in the Post-Bellum South: A Study of the Economic Impact of Racism in a Market Environment." *Journal of Economic History* 33 (March 1973), pp 131-48.

Taylor, Paul Schuster. "Plantation Agriculture in the U.S.: Seventeenth to Twentieth Centuries." *Land Economies* 30 (May 1954), pp. 141-52.

Taylor, Rosser H. "Post-Bellum Southern Rental Contracts." *Agricultural History* 17 (April 1943), pp. 120-28.

Teele, Benjamin R. "Natchez Planter Gives Up." *The Cultivator and Country Gentleman* 31 (February 1868), p. 128.

Thompson, Edgar T. "The Natural History of Agricultural Labor in the South." In David Kelly Jackson, ed. *American Studies in Honor of William Kenneth Boyd.* Durham, N.C.: Duke University Press, 1940, pp 111-74.

Trant, James B. "Financing the Production and Marketing of Cotton." *Southwestern Social Science Quarterly* 12 (June 1931), pp. 51-60.

Wagstaff, Thomas. "Call Your Old Master—'Master': Southern Political Leaders and Negro Labor During Presidential Reconstruction." *Labor History* 10 (Summer 1969).

Wiener, Jonathan M. "Planter Merchant Conflict in Reconstruction Alabama." *Past and Present* 68 (August 1975), pp 73-94.

———. "Planter Persistence and Social Change: Alabama, 1850-1870." *Interdisciplinary History* 7 (Autumn 1976), pp 235-60.

Woodman, Harold D. "Post-Civil War Southern Agriculture and the Law." *Agricultural History* 53 (January 1979), pp. 319-37.

———. "Sequel to Slavery: The New History Views the Postbellum South." *Journal of Southern History* 53 (November 1977), pp. 524-554.

Wright, Gavin, and Kunreuther, Howard. "Cotton, Corn and Risk in the Nineteenth Century." *Journal of Economic History* 35 (September 1975), pp. 526-51.

Zeichner, Oscar. "The Transition from Slave to Free Agricultural Labor in the Southern States." *Agricultural History* 13 (November 1939), p. 32-33.

Zepp, Thomas M. "On Returns to Scale and Input Substitutability in Slave Agriculture." *Explorations in Economic History* 13 (April 1976), pp. 165-78.

THESES AND DISSERTATIONS

Aikman, John D. "Mount Ararat: A Study of the Development of a Natchez Area Plantation." Ph.D. dissertation, Stephen F. Austin State College, 1963.

Breese, Donald H. "Politics in the Lower South During Presidential Reconstruction." Ph.D. dissertation, University of California at Los Angeles, 1963.

Bringer, Gladys Stella. "Transition from Slave to Free Labor in Louisiana After the Civil War." M. A. thesis, Tulane University, 1927.

Brownlee, Wilson E. "Mechanization of Southern Agriculture, 1865-1880." M.A. thesis, University of Wisconsin, 1965.

Bull, Jacqueline Page. "The General Store in the Southern Agrarian Economy from 1865 to 1910." Ph.D. dissertation, University of Kentucky, 1948.

Cox, LaWanda F. "Agricultural Labor in the United States, 1865-1900, With Special Reference to the South." Ph.D. dissertation, University of California at Berkeley, 1941.

Ellis Dorothy Lois. "The Transition from Slave Labor to Free Labor, with Special Reference to Louisiana." M. A. thesis, Louisiana State University, 1932.

Ganus, Clifton L., Jr. "The Freedmen's Bureau in Mississippi." Ph.D. dissertation, Tulane University, 1953.

Gragsby, Earl S. "The Social and Economic Aspects of Negro Farm Labor on Large Cotton Plantations, Concordia Parish, Louisiana." M. A. thesis, Louisiana State University, 1937.

Seagrave, Charles E. "The Southern Negro Agricultural Worker: 1850-1870." Ph.D. dissertation, Stanford University, 1971.

Shlomowitz, Ralph. "The Transition from Slave to Freedmen: Labor Arrangements in Southern Agriculture, 1865-1870." Ph.D. dissertation, University of Chicago, 1978.

INDEX

About the Author

RONALD L. F. DAVIS is a Professor of History at California State University, Northridge. His articles on freed blacks and sharecropping have appeared in *The Old South to the New*, *Essays in Economic and Business History*, and the *Journal of Negroe History*.